TRENDS IN JAPANESE MANAGEMENT

Also by Toyohiro Kono

STRATEGY AND STRUCTURE OF JAPANESE ENTERPRISES
LONG-RANGE PLANNING OF JAPANESE CORPORATIONS
STRATEGIC MANAGEMENT OF JAPANESE COMPANIES
TRANSFORMATION OF CORPORATE CULTURE (*with Stewart Clegg*)

Also by Stewart Clegg

FRAMEWORKS OF POWER
MODERN ORGANIZATIONS
HANDBOOK OF ORGANIZATION STUDIES (*with Cynthia Hardy and Walter R. Nord*)
CHANGING PARADIGMS: The Transformation of Management Knowledge for the 21st Century (*with Thomas Clarke*)

Trends in Japanese Management

Continuing Strengths, Current Problems and Changing Priorities

Toyohiro Kono
and
Stewart Clegg

palgrave

First published 2001 by
PALGRAVE
Houndmills, Basingstoke, Hampshire RG21 6XS and
175 Fifth Avenue, New York, N. Y. 10010
Companies and representatives throughout the world

PALGRAVE is the new global academic imprint of
St. Martin's Press LLC Scholarly and Reference Division and
Palgrave Publishers Ltd (formerly Macmillan Press Ltd).

ISBN 0-333-92970-5

This book is printed on paper suitable for recycling and made from fully managed and sustained forest sources.

A catalogue record for this book is available from the British Library.

Library of Congress Cataloging-in-Publication Data
Kono, Toyohiro.
Trends in Japanese management : continuing strengths, current problems, and changing priorities / by Toyohiro Kono and Stewart Clegg.
p. cm.
Includes bibliographical references and index.
ISBN 0-333-92970-5
1. Industrial management—Japan. 2. Manufacturing industries—Japan—Management. 3. Corporations, Japanese—Management. 4. International business enterprises—Japan—Management. 5. Strategic alliances (Business)—Japan. I. Clegg, Stewart. II. Title.

HD70.J3 K633 2001
658′.00952—dc21 00-066551

10 9 8 7 6 5 4 3 2 1
10 09 08 07 06 05 04 03 02 01

Typeset by Kolam Information Services Pvt Ltd, Pondicherry, India

Printed and bound in Great Britain by
Antony Rowe Ltd, Chippenham, Wiltshire

Contents

List of Tables

List of Figures

Preface

Japanese management is no longer the 'flavour of the month'. Its heyday in the West coincided with that moment in the early 1980s when US confidence – both political and economic – seemed to be at its lowest ebb; after the debacles of Vietnam, the Iranian hostage saga and the lacklustre Carter administration. Smokestack industries were dying, the agricultural heartlands appeared to be in great distress, and everywhere US industry seemed to be under threat. And the number one threat was clearly Japan. In VCRs, cars, cameras, entertainment systems and televisions – to name but a few – US manufacturers were rapidly losing ground to Japanese manufacturers, in some cases, such as VCRs and colour televisions, vacating the field entirely. Not surprisingly, students and professors in the business schools wanted to know how it was possible, how Japan had managed to become predominant. That was the 1980s.

Now, fast-forward the overview to the turn of the century. What a difference a decade has made. In the brave new world of e-commerce, the US economy is absolutely and unchangeably number one. The new industries of the digital decade are dominated by US software, with Microsoft as a metaphor for the new economy. And what now of Japan? It struggled through the 1990s suffering the overhang of the late 1980s bubble economy, financial collapses, political incapacity, real estate crises and loss of national industry to foreign firms such as Renault. Who is interested in Japanese management now?

We argue that it would be a great mistake to write off Japanese management. Just because the Japanese economy has experienced slow growth and the average share and real estate prices have fallen, caused mostly by speculative investment by financial institutions, the practices that made companies such as Sony, Toyota and Nintendo household names have not suddenly disappeared. It was not Japanese manufacturing firms that committed these mistakes for they have continued to hold competitive power in the world market, even though their financial performance has been affected by depression in the home market. Most of these firms remain strong and globally competitive. As evidence of their competitive power, the exchange rate of the yen remains strong: it was 300 yen per US dollar in 1976, 250 yen in 1985 and about 100 yen in 1999. The yearly trade balance in the last ten years has been approximately US$100 billion and Japan is still producing many high tech products, such as automatic cameras, DVDs, LCDs, semiconductors, small ball bearings and cars that dominate the world market.

The Japanese management system is changing but its essential strengths are not eroding. Japanese firms have a number of crucial international advantages, in comparative terms. They still have a long time horizon and make large-scale investments for the future. Synergistic, related diversification through internal development and an extreme competition focus are key characteristics. They maintain 'soft' and flexible organisational structures that, while bureaucratically administered, are not arteriosclerotic. Moreover they are based on learning capacities grounded in respect for people and for the everyday knowledge that these people possess.

In this book we detail – clearly, systematically and with considerable case and survey data drawn from leading Japanese companies – the contemporary trends, issues and problems in Japanese management. These are not 'outsider' views: we use a wealth of data and case material that only insider knowledge could produce. To add weight to the analysis, the book presents extensive data on over 200 manufacturing companies. We use this data to identify factors in the success of contemporary Japanese corporations. Of these companies, five are analysed in detail: Toyota, Sony, Matsushita, Hitachi and Canon.

This book examines not only the strength of the Japanese management system but also its problems. The contemporary trends we report are not forecasts but the new directions being taken by successful companies. Amid these new trends a number of central factors have remained the same. Balancing multiple goals while maintaining respect for employees remains of cardinal importance. Such respect has the highest priority because of the importance of knowledge accumulation in high technology environments. It is important to note that the time span attached to Japanese corporate goals appears not to be shortening under the changed economic conditions. Long-term vision remains essential to the development of high technology products and has not been sacrificed in favour of maximising short-term profits and enhancing share values.

Some contemporary trends represent a shift from the past. Horizontal strategic alliances are increasingly being developed to expand the scale of production. This is particularly important in terms of maintaining an internal development strategy, typical for Japanese corporations.

Such a strategy differs from the frequent use of acquisition and divestment in other countries, such as the US. Specialised career courses have been created so that specialists, rather than generalists as in the past, are encouraged and respected.

However some practices of the recent past have been retained, such as a wage structure based on the status ladder system. While external labour markets, pay for the job and downsizing have played a major role in countries such as the US, they are not found to a comparable degree in Japan. The employment system still places great emphasis on employability, and although the lifetime employment system has changed to some extent the big corporations still try to take care of the careers of their employees, even after retirement. While this may seem paternalistic to some foreign critics, it has led to a remarkable capacity for everyday, incremental learning and innovation at the heart of a mighty economic machine. It is certainly different from the hire and fire system, and places a very different set of knowledge management capacities at the heart of the organisation.

After 1990 many companies suffered from low profits or financial loss, and some were allowed to go bankrupt – itself divergence from the bank–industry relations that previously prevailed. However these failed companies were mostly in financial services and rarely in manufacturing. In the past some of these companies were undoubtedly too oriented towards growth as an end in itself. Too many investments were made in unrelated areas; banks made promiscuous and reckless loans based on inflated real estate values, while construction companies became mired in trying to master hotel operations. Decisions were not made by consensus but by a few autocratic leaders, on premises that were more intuitive than analytical. Bad decisions were not corrected, instead, in an attempt to recover from failure, more mistakes were made and then repeated as firms fell deeper into debt. Problem solving was put off until disaster loomed like a tsunami on the near horizon. While such problems were not due to the Japanese management system they have changed some of its features, which we shall address in this book.

The authors are experienced writers on Japanese management. Toyohiro Kono is already well-known for detailed, original research into the structure and behaviour of Japanese enterprises, as Malcolm Falkus observed in the foreword to the book that was, in some respects, the precursor of the present one – *Strategy and Structure of Japanese Enterprises* (Kono, 1984a). Toyohiro Kono began to collaborate with Stewart Clegg in the study of Japanese management when they prepared a joint book entitled *Transformations of Corporate Culture: Experiences of Japanese Enterprises* (1998). Stewart Clegg's interest in Japanese management was already well-established and among his earliest cross-national published works were *Enterprise and Management in East Asia* (Clegg *et al.*, 1986), *Capitalism in Contrasting Cultures* (Clegg *et al.*, 1990) and *Modern Organizations* (Clegg, 1990). In these works he developed theoretical considerations that he brought to the models constructed by Kono. This is the second book on management upon which they have cooperated. In the interim period Clegg, together with Thomas Clarke, wrote a book on *Changing Paradigms* (1998), which provided a new focus on some contemporary trends in Japanese management.

On this occasion the Kono–Clegg collaboration has been based, as before, on data collected by Kono and the tentative models that he constructed in Japanese. Kono's material was translated into English and became the raw material that Clegg worked on to analyse the data from a worldview that was clearly non-Japanese. He neither wished to excoriate Japan for its differences from the US model nor sought to lionise it as something culturally unique and distinctive. Thus contextually thick, deep and rich Japanese insight was blended with critical and informed analysis that enabled them to see both the specific and the general in a unique society of organisations, when regarded in a comparative context.

In writing the book there was one temptation that the authors strenuously sought to avoid. They do not see Japanese management as converging with US models. Convergence has long been a leitmotiv of comparative management studies, usually retrospectively projected from the US onto whatever area was seeking to 'catch up' with the shifting definition of modernity that its practice provided. On the contrary, the authors have sought to identify the institutional specificity of Japanese management. And they have sought to do so in ways that are neither culturally reductionist nor culturally abstracted and alienated. Just as Japanese management opened a road to more successful management systems for American firms after the first wave of interest in the 1980s, we believe that contemporary trends in Japanese management have the potential to lay a foundation for successful management systems elsewhere in the world.

This book was created with the cooperation of many people. In Japan, Professor K. Yazawa did most of the computer analysis of the data, while Miss T. Kaji and other students helped with the data collection. Mrs M. Akao produced the initial typescript. The many presentations of company cases at the monthly study meetings with corporate planners at the Japan Productivity Center were most useful. The authors wish to express their thanks to all these people.

In Sydney, Ms Cleusa Lester gave most generously of her time and organisational ability and arranged the details of Kono's visit to Sydney to work with Clegg in July 2000, when the final work on the manuscript was completed. The authors wish to thank the University of Technology, Sydney (UTS) for its support throughout the research and writing, as well as Gakushuin University in Tokyo. Their biggest thanks go to their partners – Mizu and Lynne – who managed the time to make this book happen.

TOYOHIRO KONO
STEWART CLEGG

1 The Context of Japanese Management

INTRODUCTION

In the decades before 1980 Japanese management was largely spurned for being insufficiently American (Dunphy, 1986, 1987). In the decade after 1990 it was condemned for being too Japanese. In between was the decade of the rising sun, when many books were published in praise of Japanese management. Were these books just chasing a fad? Are Japanese management and organisations still capable of continuing adaptation and change?

Books typical of the 1980s genre included *Japan as Number One* (Vogel, 1979), *Theory Z* (Ouchi, 1981) and *Made in America* (Dertouzous *et al.*, 1989). In fact the most important of them was not actually on Japanese management at all – *In Search of Excellence* by Peters and Waterman (1982). Although this book dealt with cultures of excellence in US corporations it did so in response to the 'Japanese threat'. The authors sought authentically American sources of cultural advantage in US corporations to rival the advantages of Japanese corporate cultures.

Since 1990 the growth rate of the Japanese economy has declined. Many financial institutions have had the new and unpleasant experience of dealing with fiscal crisis. The value of the land assets that were used as collateral for bank loans has diminished sharply. The world of e-commerce has flourished since the bubble economy collapsed in the early 1990s and Japan appears to be lagging behind the US in its development, although in some areas of Tokyo, such as Shibuya, the new economy is emerging. To many contemporary commentators Japan seems to be lagging both in the development of the new economy and the maintenance of the strengths of the old economy. At the turn of the century Japan is apparently a nation in decline rather than ascendancy, as in the 1980s.

In response to this decline a new wave of books and articles has appeared in criticism of Japanese management systems. Typical of these are *Japan: A Reinterpretation* (Smith, 1997), *Inside Kaisha: Demystifying Japanese Business Behavior* (Yoshimura and Anderson,

1997) and *Reinterpreting the Japanese Miracle* (Crawford, 1998). Just as the earlier studies were too one-dimensional in their resolve that Japanese management was either absolutely bad or good, depending on the decade sampled, the same may be said of the present generation of critiques. Not only are they too one-dimensional, they also assume that the future of Japanese management can be extrapolated from some of the headlines in the relatively recent past, particularly in the financial sector. We beg to differ. We maintain that the 1980s were not just a faddish moment and in this book we systematically chart the key features that will sustain the best Japanese enterprises throughout the twenty-first century.

THE SUBJECT OF ANALYSIS

Framework of the analysis

In this book we explore the strategy, structure and strategic decisions of Japanese manufacturing corporations. Other books foreshadowed our study. For instance our analysis is close in spirit to a number of works, including Rumelt's *Strategy, Structure and Economic Performance* (1974), Channon's *The Strategy and Structure of British Enterprise* (1973), Dyas and Thanheiser's *The Emerging European Enterprise* (1976), Dertouzos *et al*'s *Made in America* (1989), Fruin's *The Japanese Enterprise System* (1992) and Liker *et al*'s *Engineered in Japan* (1995). The framework of the analysis is shown in Figure 1.1. We focus on six subsystems: (1) top management and governance structures, (2) goals and philosophies of the organisation, (3) product-market strategies, (4) capability structures, (5) operations and (6) the decision-making process.

The six subsystems, as represented in the simplified model, have a definite pattern of contingent relationships. The model specifies both fitting and mismatching relationships, as well as reverse causality through feedback loops. For example, top management may choose a new product-market strategy yet may fail to anticipate the consequence of so doing. A new product mix may change the key skill sets required for the company. Whether or not managers with the new skills are promoted to top management can vitally affect the success of a strategy. The ways in which the goals of the organisation are defined by a coalition of top managers affect not only the product-market strategy but also, conversely, the goals themselves. When subsystems

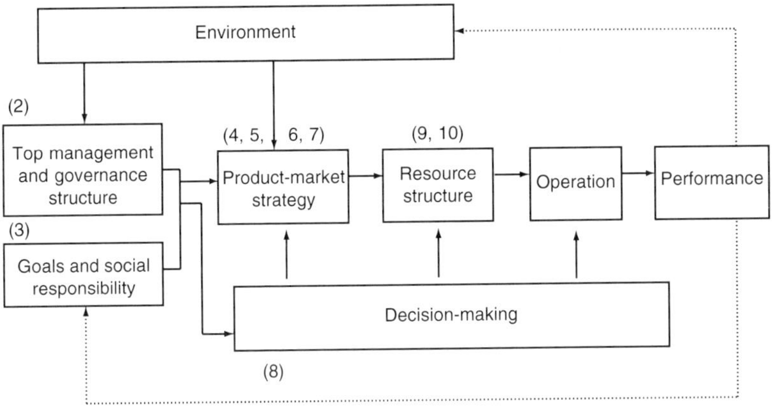

Notes: The arrows indicate the direction of relations; mutual interrelations are omitted for simplification. The numbers indicate the relevant chapters of this book.

Figure 1.1 Framework of analysis

change in a structurally adaptive way as the environment changes, then performance will improve. For such changes to flow through successfully it may be necessary to restructure the power relations. The dominant coalition presently constituted may well resist such a restructuring and if they are successful then the adaptation may be negative rather than positive – the skill set required to manage the changes will not be installed. Issues of managerial control can thus undercut product-market strategies.

Product-market strategy is a major area of organisational calculation. Should products be diversified, vertically integrated or managed multinationally? What should be the competitive strategy? Thus product-market strategy affects organisational structure. In order for operations to be carried out efficiently and for organisational performance to be high, the product-market strategy chosen must meet the needs of the market as well as be supported by the core competencies of the organisation.

Organisations selected

The research is based on a purposive sample of 205 relatively successful large manufacturing corporations. A breakdown of these 205 companies is provided in the appendix to this chapter. The firms are drawn from one of the world's most successful manufacturing industries. The advanced firms of the technology-intensive, manufactured goods sector in Japan have managed to maintain competitive power in the world market, notwithstanding the recession of the 1990s. They have done so despite the failure of businesses in other sectors, such as the service and financial sectors. Although the service sector accounts for the largest proportion of employment in Japan the manufacturing industry supports that sector. Manufacturing is Japan's key industry, enabling the country to import food, oil and other natural resources supporting the trade balance.

Of the 205 companies, six were selected for intensive analysis: Toyota Motors, Honda, Hitachi, Matsushita Electric, Sony and Canon. The companies range from highly specialised to extensively diversified. The sources of information include published materials, particularly economic and business journals, a number of visits to the companies, mailed questionnaire surveys and casebooks on the companies.

FEATURES OF JAPANESE CORPORATIONS

Long-term orientation and global vision

Successful Japanese corporations are inclined to emphasise long-term goals and have a global vision, while US corporations put more emphasis on short-term profit (Table 1.1). According to our data, about 80–90 per cent of large Japanese corporations engage in long-term planning, whereby they plan how they will grow or rationalise over a period of five to 10 years. (Our survey in 1995 showed that 93 per cent of the sample of 97 companies had a long-term plan.) They invest heavily in research and development: for example about 10 per cent of the sales is invested in R&D at Sony, Matsushita, Toshiba and Hitachi. According to our survey of 205 manufacturing companies, average R&D spending was about 4 per cent in 1995, a level consistent with the achievement of long-term growth rather than short-term profits. These companies have invested a large proportion of their resources to expand the production of technology-intensive products.

These long-term plans are reflected in the corporate vision that these companies espouse. Many of them have a clear statement of their corporate creed and vision. These emphasise the most important

Table 1.1 Features of the Japanese management system

	Japanese model (Gemeinschaft)	*Traditional US model (Gesselschaft)*	*New trends in Japan (new Gemeinschaft)*
Goals and vision	Long-term growth and global vision	Short-term profit	Balancing multiple goals of stakeholders, but respect for people has highest priority
Strategy	Competition oriented; vertical alliances	Anti-monopoly law inhibits alliances; independent company behavior	Competition and alliances
Organisational structure	Organic organisation; good interface between departments	Mechanistic organisation	Specialised career courses; moderate concentration of authority
Personnel management	Based on respect for people; learning organisation	Human resources can be bought from outside	Employability; larger differentiation of wages and promotion
Decision-making	Decision-making by consensus; sharing of goals and information	Decision-making by individuals with authority and responsibility	Initiatives and consensus

values of the corporation, for example 'Sony thrives on exploring the unknown', while Matsushita promises that 'the company will provide home appliances at a reasonable price, like the supply of water'. These long-term goals and clear corporate philosophy and vision have a great impact on corporate strategy. In part they contribute to the outstanding capacity for sustained corporate action that originates in a special conception held by Japanese corporations, a conception illustrated by the tendency of employees to speak of 'our company', meaning that the company is not only an organisation of employees but also an organisation for employees.

Restructuring is also important. Companies such as Nippon Steel, NKK and other steel manufacturers have made huge capital investments to modernise and expand their production of iron and steel, resulting in the capture of a major share of the market, even though their financial performances have fluctuated. Unlike US Steel, they have not diversified into unrelated areas because they believe that their mission is to supply essential materials to the nation. NEC, Fujitsu, Toshiba and Hitachi have made large investments in the production of semiconductors and computers, and they too have captured a significant share of the world market and experienced a fluctuating financial performance.

Japanese manufacturing corporations have to be globally oriented because they either have to export their products or establish foreign factories, although the domestic market is relatively large. Sony's policy is that all new products have to be sold worldwide and this policy is followed from the inception of these products. Overseas sales account for more than 70 per cent of Sony's total consolidated sales. Overall, Sony is more global in outlook than most, but the average firm in the sample was still globally oriented – the export ratio and overseas production ratio of the sample of manufacturing companies were 20 per cent and 16 per cent respectively in 1993.

Competitive and cooperative

Japanese corporations are competition-oriented because long-term growth and market share are important goals. Japanese corporations tend to form business groups: vertically integrated groups, and weak, cooperative, horizontal groups that are sometimes referred to as *zaibatsu*. However it is erroneous to call these groups *zaibatsu*: the original *zaibatsu* were dismantled after the Second World War and the current business groups are less tightly coupled. They are sometimes referred to as 'headless *zaibatsu*' to signify their more corporate and less patriarchally dominated structure. The top management and employees regard other companies in the group as friends, while companies in the same field but in other business groups are seen as rivals. Many companies enter into a growth field rather than stay in a niche; thus there are six major and two minor companies in the car industry, six companies (excluding foreign affiliates) in the computer industry, and five major companies in the iron and steel industry. Such concentration in the domestic market has led to intense competition and the companies that have survived enjoy a considerable competitive edge in the global market. As Porter (1990) argues, intense competition in the domestic market helps breed global champions. These successful companies are quick to enter growth areas because they are highly growth- and competition-oriented.

Japanese corporations tend to cooperate with companies that have complementary capabilities, and the development of strategic alliances has long been a popular strategy. Typically, Japanese corporations prefer to form alliances than to acquire additional companies. The first stage in their development is often to obtain a licensee agreement for a foreign technology, which they then improve. The next move is to establish alliances with suppliers and vendors – it was this that enabled the emergence of concurrent engineering and just-in-time production systems. The third move is to establish joint ventures to facilitate entry into Asian developing countries. The most recent tactic has been to develop horizontal strategic alliances to produce high-technology products, such as new product development in the pharmaceutical industry and mutual exchange of Original Equipment Manufacturers production in technology-intensive products.

Centralised and 'soft' organisations

Successful Japanese corporations are centralised organisations. Centralisation is essential to the making of large-scale strategic moves. The head office is large, research laboratories are centralised, and product divisions, where they exist, do not function as independent units. The structure of the overall organisation is not as sharply defined as in the divisional US firm. Centralisation is also beneficial to the development and consolidation of the core competencies of the corporation. While centralised, organisation tends to be 'soft' and organic, a structure in which jobs are defined ambiguously and group decision-making characterises each level of the hierarchy. The consequence is an organisation that is flexible and jobs that are easily adaptable to technological innovations.

Successful Japanese corporations respect their workers. They provide many opportunities for promotion and wage increases, and do not distinguish between blue- and white-collar workers. The lifetime employment system is not a formal contract but a commitment on the part of both the management and those employees to whom it is extended: those in the core internal labour market. Companies seek to retain such workers by providing a career with the firm, while the workers are in turn motivated to stay with the company until retirement age. It is this system that enables a Japanese company to become a learning organisation. The company can emphasise training, and employees can be rotated to gain a broad knowledge base during their long years of service.

In contrast to the Japanese pattern, in the US system human resources are bought in rather than developed internally. Workers are employed for certain jobs only, for which they have been trained elsewhere, and training is not provided by the company (Dertouzos *et al.*, 1989). Employees are easily laid off when the operation needs to downsize, and they move from one company to another in pursuit of better wages or job opportunities. In the Japanese corporation flexibility is not achieved at the expense of core employees but by means of the many part-time, seasonal and subcontracted workers that the large corporations employ and retrench as required.

Japanese corporate decisions are arrived at by consensus, and group decision-making is practiced from the bottom to the top: the management committee meets once a week to make strategic decisions (this will be elaborated in Chapter 2); at the middle management levels it is normal for many meetings to be held and for committees to be formed; and at the front line, quality circles provide opportunities for employees to offer ideas. The *ringi* system is employed widely, whereby documents on decisions are reviewed by all the sections concerned, then approved or modified, and signed or stamped by each section. This system is used to confirm group decisions or to make simple decisions of lesser importance. It leads to company-wide commitment to planned changes, with the result that implementation is not fraught with fear of failure and the process of implementation does not become an occasion for resistance to change.

The group decision-making system entails the collection of a great deal of information and ideas, and the considerion of these from many different angles in an analytical fashion. Detailed information has to be supplied in order to win over the participants, who share information and ideas. In this way companies become learning organisations and develop core competencies. Another merit of the group decision-making system is that because so much information is collected the degree of risk involved is made clear. Consequently decisions often involve more obvious risk-taking than would otherwise be the case. (The merits of such risk-taking are debatable and will be discussed in Chapter 8.) The system also ensures that responsibility is diffused. Group decision-making may be time consuming and decisions may be delayed, but implementation is faster because the participants understand not only the decisions but also the process that produced them.

The group decision-making system is highly characteristic of Japanese management practices but it is not to be found everywhere. Sometimes decision-making is the task of one individual in a position of authority and responsibility and this is reflected in the organisational layout, with individual offices for staff members rather than open-plan, communal spaces.

CONTEMPORARY TRENDS

New trends are emerging due to the changing power of stakeholders, the development of mega-competition, faster technological innovation and slower economic growth (see Figure 1.3). Some domestic commentators have referred to the emerging Japanese management system as a 'new *gemeinschaft*', premised on a new community of interest, a new type of 'stakeholder capitalism' in which several vested interests are represented (on stakeholder capitalism, see Clarke and Clegg, 1998). This differs from shareholder capitalism in that shareholders are not the only ruling power. There are three main stakeholders: employees, banks and shareholders.

Compared with the cavalier attitude towards employees in the US, as illustrated by the cut-and-slash tactics of such corporate legends as Al 'Chainsaw' Dunlap, Japanese employees are highly respected. In a high technology environment it is the creativity and knowledge of employees that creates core competencies and respect for people. However, while the expression of respect for employees in core labour market positions is hardly new in Japan (Clegg, 1990), the way in which this respect is manifest is changing. In the past it was manifest in 'service egalitarianism', in which lifetime employment was the crucial variable. Strategic employees were a part of a core labour market in which long service was rewarded by annual salary increases and in which team work and job rotation were the norm. Today the system is evolving into one where different pay for different skill levels characterises remuneration policy and where specialisation, which also boosts the employability of workers, is increasingly replacing more generalist competencies. A number of unfolding changes can be identified.

From an emphasis on growth to balancing multiple goals

The power of shareholders increased substantially during the 1990s. Because of the low-growth economy, share prices declined considerably and shareholder dissatisfaction rose accordingly – when growth was the norm, shareholders were far more complacent. Additionally, the number of foreign shareholders of stocks listed on the Tokyo Stock Exchange increased by 10 per cent on average. Overseas investors are primarily interested in the value of their shares and are much more sensitive to share prices. Hence many companies now place more emphasis on ROE (return on equity) and EVA (economic value added, net profit after interest and appropriate return to equity).

Today, Sony uses EVA as a measure of performance and it has found that EVA is related to share price. It estimates EVA for three years – thus it tries to avoid too much emphasis on measures of short-term profit. Matsushita Electric has started to use ROE as one of its corporate goals. However respect for people is a continuing value and employees remain a prime priority. For example Matsushita tried to raise the retirement age to 65 years (strictly speaking, one should refer to this as the 'extended retirement age' because employees should retire when they reach 60. Usually they enter into a new employment contract with less generous provisions because the long service obligation no longer applies). These new trends do *not* signal a convergence with the US system but rather the emergence of a hybrid model, sometimes referred to in Japan as the 'new *gemeinschaft*' model.

New forms of competitive strategy: alliances and cooperation

In the past, Japanese corporations were extremely competition oriented because of their 'me-too' or copycat behaviour. To cope with the competition, vertical alliances or *keiretsu* were used to reinforce capabilities, but recently horizontal alliances and horizontal mergers have become more commonplace. Alliances can be classified into three types: contract alliances and joint ventures based on combined strength; horizontal and vertical alliances based on the different skills offered by the partners; and alliances where different development, production or marketing processes are combined. For example Hitachi and NEC, once rivals in the semiconductor business, decided to cooperate in the development of large-memory semiconductors, as did Toshiba and Fujitsu. Likewise Nippon Steel and Sumitomo Steel are cooperatively producing H-type steel products, while Nippon Oil merged with Mitsubishi Oil and the new company is cooperating with Cosmo Oil with regard to sales channels. The are several reasons for the recent increase in horizontal alliances, including the rise in development costs and the increasing importance of economies of scale. Furthermore, compared with mergers and acquisitions, alliances are more flexible and may help ward off foreign intervention while intensifying competition with other groups, thus contributing to overall competitiveness.

Alliances are not new in Japan. They have been popular for many years because of the group-oriented nature of firms and have often served as a substitute for mergers or acquisitions. Nonetheless, according to Thompson Financial Securities data, mergers and acquisitions soared by 340 per cent in 1999 to a record eight trillion yen (*Australian Financial Review*, 10 January 2000, pp. 14–15). Of these, nearly a third involved foreign companies, whose participation in the market grew to 32 per cent of all transactions during that year – a threefold increase.

Changing conception of the career path

In organic organisations, job scope is wide and in principle people are willing to do what others expect of them. These traditional practices have not changed in Japan, but the generalist career profile, that for so long sustained and defined them is shifting. Many successful companies, such as Matsushita and Toyota, have established specialised career paths. Toyota has six specialised career groups, including general management, marketing, development, production technology, production planning and new businesses, and within these groups there are more than five career paths. More specialised paths are now considered necessary for the accumulation of technological knowledge and the building up of core competencies. Specialised career paths differ from the pay-for-the-job system and do not equate with narrow career specialisation. There are no job titles and wages are aligned with status, which is determined by skill level. This is different from the hire-and-fire system because it is based on features of the traditional internal labour market system (this will be discussed in Chapter 10).

Relaxation of centralised authority

In the past the Japanese organisational structure tended to be highly centralised but recently a degree of decentralisation has occurred. One can see this in the tendency of many companies to establish 'internal companies'. Hitachi has established ten such companies, each headed by a president. Sony has set up four companies of this type. Hitachi previously had a strong head office that housed marketing and other key functions. The many factory profit centres reported to the head office, but because they were shortsighted and afraid of failure, risky strategic decisions tended to be postponed and this hampered innovation. Sony had a hybrid organisational structure with many product divisions, but these lacked marketing and research functions because many of the key functions were centralised in the head office. Sony then established ten internal 'companies', but marketing and research were still conducted at head office. More recently Sony established four internal companies with full-responsibility for these functions. The head office still retains staff teams to study strategic issues.

The reasons for this change to the organisational structure are as follows. First, rapid strategic decisions have become necessary in the age of mega-competition: competitive advantage depends on seizing new opportunities as they appear. For instance Sony was quick to enter the entertainment and Internet businesses. Second, the construction of core competencies requires the concentration of resources into growth areas. Small product divisions with responsibility for short-term profits are not equipped to take decisive strategic action.

The shift from lifetime employment to interorganisational mobility

In order to avoid the need to discharge employees because the jobs they do have become redundant, employees' skills are updated to meet the organisation's new requirements. These updated skills are also needed by other organisation, and while interorganisational mobility was restricted in the past, companies are increasingly helping employees to move to other firms.

Tokyo Gas (12000 employees in 1999) offers six 'second life paths' to employees aged 50. The options include taking early retirement or finding a new job, working shorter hours, taking standard retirement at the age of 60, signing a new, obligation-free contract with the company at the age of 60, work-sharing after 60, or permanent secondment to an affiliated company. A counselling service is available to help employees decide which path to follow. These new conditions of employment represent modification of lifetime employment, not a contradiction of it.

When ownership has becomes vested in a foreign company the changes can be more radical. For instance Renault now controls Nissan and is seeking to lay off 21 000 employees. While the company is trying to do so in terms of a new contractual understanding based on employability, it is insisting that employees be prepared to move geographically to take a job – something that is anathema for those used to Nissan's old ways.

Larger differentiation of wages and promotion in the status ladder system

The status ladder system continues to exist: employees are promoted up the status ladder and wages are determined by status grade, not by job grade. The higher ranks climb both the job and the status ladder within the internal labour market system. However when the Japanese economy began to slow the old system of annual wage increases with small differentials had a long-term escalatory affect, increasing the labour costs as the workforce aged; thus many companies started to change their wage system.

At Toyota, wages are composed of 30 per cent basic pay, which reflects the employee's status grade, and 70 per cent merit pay, which reflects performance and ability. Wages can fall as well as rise, depending on merit achievement. The trend is towards a larger differentiation between grades while retaining lifetime employment for the internal labour market.

Increased scope for initiative and consensus decision-making

Decision-making by consensus has not changed and, as in the past, new ideas and innovations are welcomed. For many years, since the immediate postwar influence of US specialists such as Homer Sorrenson, Japanese firms have encouraged quality circle activities and suggestion schemes, and these have done much to improve the quality of jobs and lift morale. In order to ensure a creative and innovative organisation, top management not only requires innovative vision, it also needs to change the management-by-objective system and redefine the existing appraisal system. Many successful companies have recently changed their appraisal system, so that 'strategy construction' and related conceptual skills such as 'issue finding' and 'new task creation' now have more importance placed on them in the management by objectives and appraisal systems.

DIFFERENCES IN MANAGEMENT SYSTEMS: THREE MODELS

There is a common misconception that Japanese management is unique, based on a unique cultural background, and that it cannot be universalised. Yet most of the characteristics of the Japanese management style were formed after the Second World War. Japanese management was a product of rational thinking; some of it introduced from the US during the occupation, some of it developed from earlier imports such as Taylorism, moulded to the paternalist and cooperative traditions already established in Japan (Tsutsui, 1998). Many of the theories and business practices (such as management committees and quality controls) that have been seen as uniquely Japanese were in fact transplanted from the US or Europe. For instance scientific management has been seen by some commentators as the rational core of almost all Japanese management innovations from the interwar period onwards (ibid.) Japanese management is not some outcrop of a distinctive cultural universe or a relic of the feudal past. Japanese managers eagerly learnt US theories and implemented them, initially under US tutelage during the occupation but freely thereafter. As more than one American professor of management has remarked to us, 'why didn't American managers learn our theories when the Japanese did?'

Within management there are three approaches, broadly conceived, which seek to address issues of cross-cultural adaptation, isomorphism and distinctiveness in management systems. Views differ, with some theorists seeing management as determined by culture or technology, while others argue that determinism is not the appropriate concept for thinking about management. As a socially constructed phenomenon, one that is produced by people from the various pieces of knowledge that enter into their thinking and practice, it is more appropriate to see management as the result of strategic choices made by key actors and institutions. The latter is our view, but we shall first discuss the determinist position.

Cultural determinism

The emphasis here is on how a specific social culture and its value system determine the features of a management system. Max Weber (1904) maintained that an 'elective affinity' existed between the ideology of early American capitalism and the strictly individualist accounting principles of Protestantism. In a similar vein, Hazama (1963) insists that the traditional concept of family life determined the features of Japanese companies. He regards a corporate workforce as comprising members of a company family, and that the traditional concept of family was the origin of lifetime employment. When we use a social model to explain the group-oriented nature of Japanese corporations and their members, we are employing a culturally determinist theory (Benedict, 1946; Nakane, 1967; Doi, 1971, also develops a logic of *Amae* or reliance). Reliance on the group is seen to entail a long-term commitment to the same employer. The traditional respect for elders in Japan is also sometimes used to explain the wage and promotion system based on length of service. The emphasis is on management systems being determined by the social culture.

The majority of features of the Japanese management system cannot be explained by this approach. Before the war, showing respect for people and offering lifetime employment were not popular practices. Indeed the culture was extremely militaristic and authoritarian as Japan was preparing for war and domination of the East Asian Greater Co-Prosperity Sphere. Military models were paramount. There was a great difference in status between white-collar and blue-collar workers. White-collar employees could leave the office at noon on Saturdays but blue-collar employees had to work long hours each day, with ten hours not being unusual. The working conditions on fishing boats and in silk factories and coalmines were particularly bad, and were often arenas for left-wing agitatators, who, despite fierce repression in the 1930s, could still be heard from time to time. In many respects, since the 1930s the values of militaristic Japan have been subdued (although they have not been entirely eliminated) and management systems have changed greatly, although many other aspects of traditional Japanese social values remain popular and consistent with the past. That social values have stayed almost the same while the management system has changed is a serious drawback to explanations from this approach.

However some useful questions can still be addressed from this perspective. For instance, which aspects of its management system can Japan export? Reduced job specialisation and long-term commitment to the same employer might be exportable to countries where individualism is more prevalent than in Japan, but with limitations. Similarly there are limits to what can be imported to Japan: job classifications and job-related pay were imported around the 1960s because they were 'rational', but they did not function well. All the companies that adopted these practices eventually abandoned them and returned to the length-of-service system because the cultural sense of fairness had not changed. Thus social values can affect management systems.

Technological determinism

In this approach it is the production technology that determines the management system, as in Marx's theory that the property relations of a production system determine the political and social system (Marx and Engel, 1848). In the field of management, Greiner's five stages of development model also emphasises technological determinism (Greiner, 1972). As the company grows, products are diversified and as the product-division structure expands more rules are needed to control the operation. Such models tend to take a convergence approach, assuming that all management systems will be 'modernised' and thus converge into something like the American system (Marsh and Mannari, 1976). This model argues that when a company diversifies into a technology-related product mix it will tend to have a large head office, and when a company enters into unrelated diversification by acquisition and divestment it will tend to have a small head office.

In Japan the long-term employment system first emerged in heavy industry – for example the machinery, shipbuilding and iron and steel industries – because skilled labour was required. In order to retain their labourers these industries had to provide job security. After the war, when the industrial structure shifted even more towards heavy industry employing skilled labour, the lifetime employment system became more widespread, in large part because of union demands in the period of postwar liberalisation.

Technological determinism seeks a general or uniform theory applicable to any country. It is similar to convergence theory in this respect but it attributes the effect of convergence to some underlying technological force. Other convergence models stress alternative factors. For instance a very influential UK model, 'late development theory', argues that Western management systems will converge with the Japanese system (Dore, 1973). In the US, Ouchi's (1981) 'Theory Z' maintains that the American and Japanese models will converge.

Strategic choice model

In this model the management system is seen as the result of choices made by top management when they enter into strategic relationships in order to find the best means of achieving strategic goals. Kono (1984a) maintains that most of the features of the Japanese management system seem to be selected through rational choices made in order to attain goals, a view with which Fruin (1992) agrees. From this perspective we would suggest that a clear statement by top management of a corporate creed would produce strategic decisions with greater integrity. A business group would then be more effective in procuring resources, and particularly in securing investment funds from its main bank, which typically will have a long-term horizon rather than a short-term orientation toward profits in the next quarter. While this may seem laudable, and at its peak in the 1970s and 1980s it was very successful, the use-by date of fostering long-term planning, accelerated investment and rapid innovation is clearly past. The best-managed Japanese organisations realise this and this realisation is behind the many subtle changes we chart in this book. While their strategic choices made sense in the climate that prevailed in the past, they did not after the bubble economy collapsed. The need for a change of paradigms was evident (Clarke and Clegg, 1998).

The late 1990s ushered in a period of readjustment around changed paradigms. At the heart of these stood the old social contract based on paternalistic corporate social welfare – a contract that never included all Japanese citizens, only those who were fortunate to be employed in the core internal labour markets of the big-name companies. At the heart of this was the lifetime employment system. Lifetime employment was adopted to cater to union demands for job security when anti-union legislation was introduced in 1946 in response to the wave of strikes and sit-ins that greeted the initial liberalisation of industrial relations by the US occupying power. Subsequently, in an unanticipated consequence, lifetime employment and the internal labour market became major means of sustaining knowledge accumulation and enhanced product quality. Intensive training programmes for employees became rational under lifetime employment as employers could be sure that the firm would reap the benefits, rather than their being poached by a competitor. Such training was one of the most important factors in the enhancement of product quality and productivity (see Dunphy, 1986, 1987, on the applicability of different models to Japan).

A simplified model of cause and effect relationships is shown in Figure 1.2. Strategic choice is the most important element. For example the rapid reindustrialisation of the postwar period, after the devastation of the war-based economy, required skilled labourers. Lifetime employment encouraged skill formation and the training costs were invested in secure personnel, people who would remain with the firm. The decision to develop a life-long learning system also encouraged respect for people, and the importance of long-term growth to organisations helped to foster this. The organic organisational structure

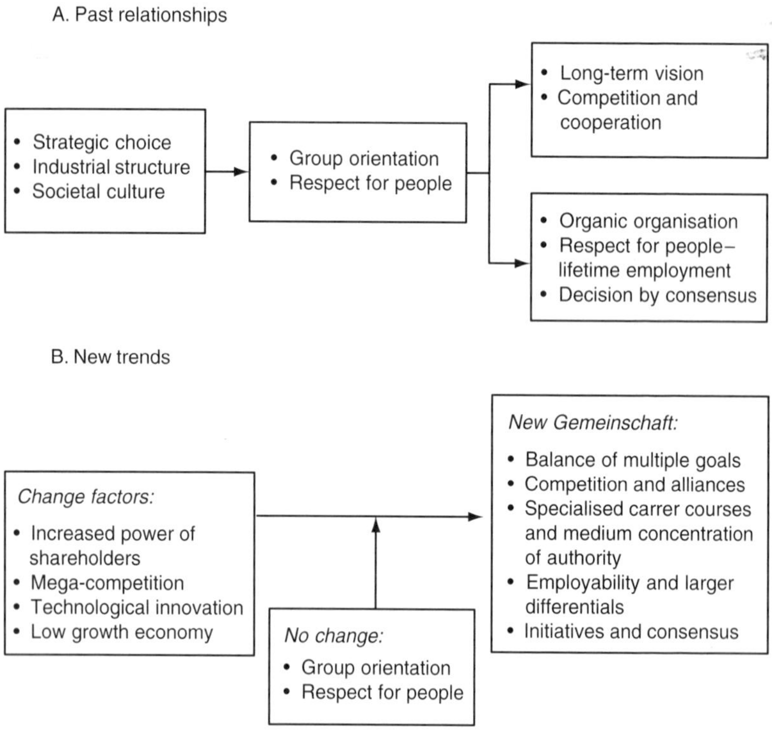

Figure 1.2 Simplified cause and effect relationships

	Universal approach	Contingency approach	Emphasis on difference
In one country	(1) Universal theory	(2) (National) contingency theory	(5) 'It all depends'
In more than one country	(3) Universal theory (or international grouping of universal or convergence theory)	(4) International grouping of contingency theory	(6) Emphasis on differences

Figure 1.3 Contingency theory matrix

worked well under the long-term employment system. Long-term employment also fostered a sense of commitment. Secure in their financial stakeholding and not subject to the threat of hostile takeover, Japanese strategic managers were free to plan more rationally over a longer time horizon and could afford to be less obsessed with their bottom-line results in the next quarter. Our way of making theoretical sense of the questions we consider is represented in cells 4 and 6 of Figure 1.3.

CRITICISMS OF THE JAPANESE MANAGEMENT SYSTEM

In the past some specific features of the Japanese management system were seen as principles that other countries could learn from. However, many books and articles began to criticise Japanese management practices as the economy slowed and financial institutions experienced very public problems. Table 1.2 lists the major problems. In this section we present some standard representations of the problems with Japanese enterprises, followed by a demonstration of the extent to which elements of the paradigm may be changing.

The value system

Social values

Japanese government policy has typically put more emphasis on production than on consumer welfare. The slogan promulgated by the

Table 1.2 Problems with the Japanese management system

	Japanese system	*Traditional US system*
Value system		
Social values	Emphasis on production, neglect of consumers	Emphasis on consumers
Corporate goals	Neglect of shareholder value, importance placed on employees	Emphasis on shareholder value and short-term profits
Strategy and structure		
Government and business	Alliance of government and business, no real competition	Industry antagonistic towards the government
Competition	Group-oriented and follower behaviour	Niche seeking
Organisational structure	Centralised power	Decentralised power
Personnel management	Enforcement of hard work	Balancing work and private life
Decision-making	Imitative (or catching up) and incremental, orderly and obedient	Seeking home-runs and opportunities for self-expression

government before the Second World War was 'Rich country, strong army'. Not surprisingly this slogan was abandoned after the war. In order to rebuild the war-torn economy, emphasis was put on production rather than consumption, a policy that continued long after Japan became the second most significant economic power in the world. The increased production led to improved living standards – the orientation towards production was not intended to mean the sacrifice of living standards, rather it was meant to improve the lives of the people. Japan's per capita GNP became the highest in the world (Table 1.3 lists some relevant comparative data on economic performance). What were the causes of this policy tropism?

The government's postwar investment policy emphasised the development of industrial land, road construction and hydro-electric schemes, rather than the construction of housing or the remodelling of cities and residential areas. A high savings rate was encouraged by the tax system and the savings to income ratio was very high (15 per cent, compared with less than 5.6 per cent in the the US). One consequence of this was that consumption remained low and hence there was great pressure to export, resulting in a huge positive trade bal-

Table 1.3 Economic and social performance, selected countries

	Japan	*US*	*UK*	*Germany*
Real growth rate (1993–7, %)	1.4	2.7	2.9	1.4
GNP per capita (1997, US$)	33.319	30.160	21.916	25.640
Unemployment (1997, %)	3.4	4.9	6.9	11.4
Trade balance (1992–7, billion $)	122	–104	–20	46
Exchange rate (national currency/ dollar)				
1977	Y300	$1.0	£0.57	DM2.32
1985	Y260	$1.0	£0.65	DM2.55
1997	Y120	$1.0	£0.59	DM1.72
R&D/GNP (1996, %)	2.93	2.55	2.05	2.29
Income distribution (%, 1996)				
Highest 10%	20.5	23.7	22.9	n.a.
Lowest 10%	3.8	1.9	2.5	n.a.
Average life expectancy (1995, years)				
Male	76.9	73.4	74.5	73.4
Female	82.9	80.1	79.8	79.9
Crime rate (1995, per 100 000 people)	1600	5200	9100	8000
(theft, homicide, arson, rape etc.)				

Sources: Keizai Koho Center, *Japan 1999, An International Comparison* (Tokyo: Keizai Koho Centre, 1999);. *Japan Almanac* (Tokyo: Asahi Newspaper, 1998).

ance.This bred very successful export-oriented companies. In the social domain, government and corporate policies that emphasised social conformity were supported by and conformed with the values held by employees and stressed by employers. Employees became long-term beneficiaries of their companies' export success and grew increasingly dedicated to the company, which provided them with security and a steadily increasing income, although sometimes to the cost of their private life. It was an exceptional postwar recovery.

The personal costs are well known. Overcrowded commuter trains hurtled salaried employees over great distances to their offices. An office culture developed that rewarded excessive zeal and long hours at work. Hotels consisting purely of small sleeping capsules were developed close to offices so that businessmen could take a brief respite from their work without having to return to their homes in the suburbs. Meanwhile family life registered the toll. Little meaningful time was spent outside the company; life was so work-focused for salaried employees that after retirement they had no friends and nothing to do each day. This pattern of life, reinforced by company policy, was very different from that in other societies, where people had enough private time to enjoy family life, go to church, make friendships outside the company and take up hobbies such as gardening.

Discussion While there is nothing intrinsically wrong with working hard (and we are confident that American, Australian or European managers work just as hard as Japanese managers), maintaining a balance between company life and private life is important. Perhaps the lack of balance in this respect is one reason why the value system of Japanese workers is gradually changing. Japanese working hours fell from 2052 hours a year in 1990 to 1919 hours a year in 1996. In 1996 the comparable hours were 1996 in the US and only 1550 in Germany (Ministry of Labor). Overseas travel from Japan increased from 5 000 000 trips in 1985 to 16 000 000 in 1996 (Prime Minister's Office), hence many more Japanese people were beginning to see what a society founded on different values might be like.

Corporate values

Japanese managers refer to the company as 'our company' rather than 'your company' (that is, belonging to someone other than themselves). By this they mean that the company exists for the benefit of employees. This is to a large extent correct: companies explore opportunities for growth at low profit rates because this type of growth increases the number of available positions for employees. One consequence of this strategy is a low return on equity (ROE). For example in 1996, the average ROE (net profit after tax divided by equity capital) of Dow Chemical, 3M, P&G, Eli Lilly, IBM, Motorola, Chrysler and GE was 23.8 per cent while that for corresponding Japanese companies such as Mitsubishi Chemical, Nintendo, Shiseido, Takeda, NEC, Fujitsu, Hitachi, Nissan and Toshiba was 4.99 per cent. The stock market (Nikkei major stock price average) declined from 38 900 yen in 1989 to 14 000–18 000 yen in 1999. If everything in the economy is deflating simultaneously, such a decline may be manageable. However the low ROE, together with the serious stock price decline, became a serious problem when the percentage of foreign shareholding began to increase. While the average ratio was 9.8 per cent for all companies listed on Tokyo Stock Exchange: 1996, for major international companies such as Sony and Canon it was more than 30 per cent. Such declines in value are more serious for foreign owners because of the cost of currency conversion.

The crisis faced by many banks and other financial institutions after 1997 derived from Japan's expansion-oriented goals. Before 1997 these financial institutions funded investors on the basis of their real estate equity, based on current land values, but the price of land dropped in 1997 to one fourth of the price prevailing in 1990. Bad debts arose partly because of expansion-oriented decisions and partly because of the government's assurance that the banks would never be allowed to go bankrupt – this in order to protect depositors. Hence the number of bad debts grew, and it was when the financial institutions started to call in entitlements in order to cover these debts that the crisis occurred.

Discussion In the past, although Japanese companies placed less importance on share prices they did not neglect profits. They tried to balance the multiple goals of profit, growth and stability in the interests of both shareholders and employees. (Indeed intense competition, aggressive investment in R&D and substantial capital investment – in semiconductors, for instance – helped Japanese corporations to attain the largest share of the world market for high technology products). Normally this would not have presented a problem, but it did so when the financial services sector and manufacturing became out of kilter with each other.

Strategy

The alliance between government and business has led to a lack of real competition. In many areas of business in Japan a license has to be obtained from a government ministry to start up a business. For example the capital investment required for steel and oil refineries, new bus lines and bank branches are all subject to licensing. The ministries granted more than 10 000 licenses in 1996 (*Asahi Almanac*, 1998). These licenses act as entry barriers not only to foreign companies but also to domestic companies, hence they are anticompetitive and effectively reduce competition. While there are anti-monopoly laws, many exceptions are granted. For example, cartels are allowed in many industries, while retail price maintenance by manufacturers is allowed in others, such as the cosmetics industry and book publishing. The law does not regard the quasi-vertical integration of suppliers and sales channels as an impediment to competition. The most powerful Ministry, MITI (Ministry of International Trade and Industry), encourages cooperation between large companies with respect to capital investment and research activities. For instance it initiated a large-scale semiconductor research cooperative, consisting of five big companies and supported by national funding. The competitive power of Japanese semiconductors in the world market was enhanced by this move. MITI has also developed long-range industrial forecasts, as well as guidelines that can be used by companies in the same industry to coordinate their capital investment, thus reducing the competition among these companies. Furthermore the government does not allow foreign companies to bid for large public construction projects, such as road and bridge building. Such public works account for about 8 per cent of GNP. Also, the purchase of computers by public organisations, including universities and research institutions, is limited to domestic products.

Discussion: Many members of the Diet have affiliations with a particular industry, from which they gain support at election time and receive financial backing. Not surprisingly, reciprocal favours are expected and they tend to support their affiliated industry with respect to government funding. This is particularly the case with the construction industry and the agricultural sector, and public investment in these areas tends to be abundant. By contrast the more recently developed information industry does not have affiliated members so there is less public spending on information systems.

Comment Government–industry affiliations were initially useful in directing and supporting weak infant industries. However they impeded competition and created entry barriers to foreign companies. Commentators in many countries once expressed their admiration of government–business cooperation in Japan, particularly the guidance exercised by MITI. Indeed this policy was particularly effective in helping Japan recover from the destruction of the war and in supporting the development of industry. However the current trend in Japan, and in nearly every country, is towards increased deregulation.

With regard to competition in contemporary Japan one should distinguish between two sectors. One is the competition-intensive sector, organised under conditions of domestic oligopoly and some government coordination. Most of the high technology producers and export producers belong to this sector, and there is fierce international competition in respect of product innovation by a small number of firms that are very large players in the domestic economy. The other sector is much more regulated and protected and includes agriculture, finance, services, small retail businesses and manufacturers of non-export goods. Productivity in this sector is much lower than the world standard.

Group orientation and the follower ('me-too') strategy

Many Japanese companies are followers, pursuing the 'me-too' strategy, including about five companies in the car industry, plus the motorcycle, computer, semiconductor and colour television industries. Many companies that belong to different business groups but function in the same product area follow a similar strategy. For example the electrical products companies in Mitsubishi, Mitsui, Sumitomo and Yasuda groups have a similar product mix, while similar companies in other countries tend to adopt niche-seeking strategies. Intel and TI produce a different range of products from each other, for instance, and VW's range of cars differs from those of Daimler-Benz and BMW. The follower strategy was originated by the *zaibatsu* and other business groupings because each wanted to have a set of growth products, so they imitated the strategies of each other.

In addition to the *zaibatsu* groups there are *keiretsu*, or vertically integrated groups where companies in the same group mutually trade their products and support each other. For example Sumitomo Bank supported fellow group member Asahi Brewery when its market share of beer fell to 9 per cent. Such a grouping tends to result in member firms being like a 'convoy of transport ships', whose speed is determined by the slowest boat but the weakest boats can survive as they are protected by the stronger ones.

Discussion The degree of solidarity within *zaibatsu* groups has been exaggerated by many non-Japanese researchers: typically there is no controlling centre, only regular meetings between the presidents; the extent of trading with companies within the group tends to be much less than with companies outside the group; and manufacturing companies within the group borrow money from non-member banks. The thesis that the groups involve tight bank control or hegemony over affiliates is not easy to sustain.

A strategic alliance is different from a *zaibatsu* grouping because the former involves cooperation between complementary rather than affiliated firms, such as joint research in the pharmaceutical industry or the exchange of OEMs in copier production. Such cooperation is becoming increasingly popular because on their own few if any companies are able fully to develop the knowledge and applications required for high technology products. Vertical alliances through the supply chain are attracting many followers in other countries because they enable companies to carry out concurrent engineering and practice agile management.

The follower (or 'me-too') strategy entails intensive competition in growth areas, and if a company can survive domestic competition it can achieve competitive strength in the world market.

Organisational structures – centralised decision-making power

Strategic decision-making is centralised in Japan but the implementation of operational decisions is decentralised. The style of personnel relations tends to be authoritative, with the senior employee having strong control over his subordinates. Under the lifetime employment system senior employees are powerful because subordinates cannot move to another company, even if they feel that a superior is being unfair. Subordinates calls their bosses by the title of the position they hold, not by their given name. Employees are required to stay late at the office if their chief is still working, and they have to work during their holidays if so ordered. They never complain about their boss to a higher authority.

Comment The question of the power hierarchy can be approached from two directions. One is in terms of the personnel management system. There is no differential treatment of white-collar and blue-collar workers. The same wage and promotion systems are applied across both groups and everybody can be promoted up the status ladder. The departmental offices are open plan and even the chief of department's desk is located in the communal office. Everybody eats in the same canteen. On the manufacturing front, all staff members and workers wear the same uniform suits and managers visit the shop floor quite frequently.

The second approach involves decision-making style. In Japan, decisions are reached by consensus. Although strategic decision-making is centralised, operational decisions are made at the lower levels. There are many meetings, quality circles and suggestion systems and these give lower-level employees a voice in decision-making. As the lifetime employment system protects the status of employees they can feel confident about stating opinions that differ from those of their superiors (see Kono and Clegg, 1998).

Personnel management: hard work and temporary workers

Respect for others is only a superficial principle in Japanese workplaces and in practice hard work is expected from all. If employees do not work hard they are transferred to meaningless jobs (*madogiwa-zoku*), and the lifetime employment system means that they cannot move to another company without causing themselves serious disadvantages. Thus white-collar workers stay late at the office and their private life is sacrificed. A popular saying is 'My daytime working steals my night home life'.

The size of the workforce is often boosted by temporary, part-time, seasonal, contract or transferred employees. In some companies, temporary employees account for more than 30 per cent of the workforce. On average, in 1996 part-time employment accounted for 36 per cent of female employees and 11 per cent of male employees, a higher ratio than in any other OECD country.

On the production front, the just-in-time system enforces hard work because there is no time to spare, no slack moments in the working day and workers on the production line never seem to stop. The *kaizen* system, which relies on the 'participation' of employees, is often subject to criticism in Japan because it puts pressure on employees to attain cost reduction targets. Full participation is rarely left to chance and is systematically structured so that it becomes the norm.

Discussion: there are two promotion ladders: the status ladder and the job ladder. Wages are determined by status and there is a wide range of wages within each status designation. Wages increase every year – only in small increments but even this can act as a strong incentive (as we shall elaborate in Chapter 10).

Companies have many temporary employees and their number fluctuates. During the 1980s there was a shortage of labour and the percentage went down, but after 1990 it started to rise again. The key point is the job security given to permanent employees – they are rarely laid off, unlike in the US, where lay-offs are frequent and staff turnover is high.

Product development: incremental and imitative

The aim of the postwar Japanese manufacturing industries was to catch up with the West with respect to product design, product quality and production technology by means of imitatation. However in doing so they usually improved and reduced the cost of the products considerably – so much so that with many products, such as videos, colour television tubes and fax machines, they obtained a stranglehold on global production, delivering what the global market wanted at prices the market wanted to pay.

Discussion: At issue here is whether production is still imitative, a subject to which we shall return in Chapter 8. A common criticism is that most of the original patents were registered in the US or Europe. While this is the case, one has to acknowledge the product developments and mass production adaptations carried out by Japanese corporations such as Canon, an outstanding company with a good growth record in the 1990s whose main products are 3.5 mm cameras, photocopiers and computer printers. While the basic technologies for these products were all imported and the patents of products such as colour televisions, VCRs and semiconductors were of Western origin, today imitation is history and many successful companies are pouring a large amount of money into research and development. Furthermore the number of Japanese patents listed with the US Patent Office is increasing. In 1998, of the 10 companies with the largest number of US patent acquisitions, six were Japanese. Many new products have been pioneered by Japanese companies, including automatic cameras, 8 mm video cameras, fax machines and large-memory semiconductors. Japanese corporations carry out both incremental improvements and innovative product development, sometimes very aggressively, as evidenced by the huge capital investments in the production of steel, semiconductors and cars.

Nonetheless Japanese product development tends to be incremental rather than innovative. Making photocopiers both smaller and cheaper is a case in point. In the case of production, the JIT system and quality refinement by quality control (QC) circles are further examples.

Employees have also adopted what one might term a follower style, as illustrated by the expression 'the nail that sticks up is driven down', meaning that group acceptance is considered more important than expressing one's own opinion. Yet these stereotypes fail to capture a system in which employee initiatives form the basis of creativity in the organisation. There is a misconception that the Japanese do not express their opinions in meetings, will not be negative, will not say 'no', but this depends on the situation. At company meetings, in QC circle meetings and through the suggestion system they do express their opinions, but may be less inclined to do so at meetings outside the company. Products with worldwide competitive strength cannot be produced without heated discussion.

THE WIDER SOCIAL, POLITICAL AND ECONOMIC ENVIRONMENT

While there are undoubtedly local aspects to consider, there are more similarities between Japan and the Western countries than there are differences. Broadly speaking, the Japanese education system is similar to that of the US and Japan's political system is similar to that of the UK. Japan specifically constructed its systems in conformity with these countries (Westney, 1987). There is a difference, however, between their business environments.

The socio-institutional environment

The demographic context

The population of Japan is 126 million, approximately twice that of the UK, Germany and France, while the standard of living is about the same as in these countries. The Japanese domestic market is large enough to ensure that new products that survive the competition in Japan will be equally competitive in the world market. Japanese consumers are very strict about the quality of the products they buy and this encourages manufacturers to strive for the highest quality.

Japan is an ethnically homogeneous country, where everyone speaks one language, people are well-educated and even the lowest in the hierarchy of workers are skilled. Life expectancy is the highest among developed countries, as shown in Table 1.3. As in many other countries the birth rate is falling (it was 1.38 in 1998) and this is causing an increase in the average age of the population. Persons aged 65 and over accounted for 4.9 per cent of the population in 1950 but by 1995 the percentage had increased to 14.6 and by 2020 it is expected to rise to more than 30 per cent. This might harm the vitality of the Japanese society. The low birth rate is due to delayed marriage, the increasing number of working women and inadequate maternity leave.

The educational context

In 1998 there were 576 universities and 2.6 million students (excluding the 598 two-year colleges, which had 0.5 million students). Approximately 350 000 university graduates and 150 000 college graduates are supplied every year to various organisations. It is not unusual for university graduates to account for more than 30 per cent of total employees in high technology companies. The proportion of young people in higher education, including two-year colleges, is 46 per cent, the same as in the US. In Germany it is 32 per cent (Ministry of Education, 1996). The increase in the number of university students was mostly made possible by the expansion of private universities. Unlike in Europe there are many private universities and the system is flexible enough to meet the demands of society.

Throughout the six years of elementary school, three years of junior high school and three years of senior high school a structured education with a uniform curriculum for every pupil is stressed, all students being considered of equal capability. The equal treatment and group training during elementary and secondary education contributes to the homogeneity of the Japanese people. Up to junior high school level a morning meeting is held every day and the school head delivers a speech. Hence the morning meeting at business enterprises is nothing new to Japanese workers. School excursions are compulsory. Many schools have school mottoes and songs. Group spirit rather than individualism is emphasised in this education system.

The values context

Devotion to the organisation and hard work are fundamental values of the Japanese people. Workers select one formal organisation and devote themselves to it alone. Independence, freedom and leisure time have not traditionally been highly valued by Japanese people. This orientation towards the organisation comes from traditional culture, which Benedict (1946) describes as a culture of shame (as opposed to a culture of sin), while Nakane (1967) writes of a vertical society versus a horizontal society and Doi (1971) refers to the Japanese people's 'tendency to depend on others' rather than seek independence.

The political context

Since the end of the Second World War the government benches have been dominated by the conservative Liberal Democratic Party (the Socialist Party has obtained a majority only once – in 1949), enabling a long-term economic policy to be worked out. However there are many factions within the Liberal Democratic Party that compete with and criticise one another and this has helped maintain the viability of the opposition parties. Because the Liberal Democrats lost a number of seats in both houses in the 1990s they now have to collaborate and compromise with other liberal parties, which has tended to delay political decisions. The Socialist Party remains a minority party.

The power of the prime minister is not as strong as that of the president of the US, who has a staff of more than 500 and appoints more than 3000 government officers. The Japanese prime minister has only 10 staff members, sent from various ministries. The prime minister and other ministers depend on information and ideas from the bureaucrats, which also serves to delay political decisions. The civil service in Japan is competent and plays a very important role in collecting information, enacting laws and running the government. Hence Japanese politicians rely on the civil service to a much greater extent than is the case in the US and the UK. In the US politicians depend on their own staff and lobby group, for information. In the UK politicians rely to a certain extent on the advice of civil servants but they also have other sources of advice, such as think-tanks.

The bureaucratic context

MITI plays an important role in the coordination of business activities, although perhaps not to the extent that some observers have suggested. MITI does not have strong legal powers but it does enjoy considerable influence. For example, when there was overcapacity in the shipbuilding industry MITI formed a committee with representatives from all the shipbuilding companies and encouraged them to work out plans to reduce their capacity. MITI has been criticised because it has created entry barriers through the imposition of permits, licences and safety rules (as in the car industry), and through its procurement policy for public organisation. This situation is now changing because of deregulation and a more liberal policy stance.

The Japanese government is not expensive to run. In 1995 public servants in central and local government offices (excluding military personnel, post office and other public employees) only amounted to 3.9 per cent of the total population, compared with 8.0 per cent in the US and 8.7 per cent in the UK. The tax and social security costs of running Japan amounted to 35.4 per cent of GNP in 1993, compared with 35.2 per cent in the US, 44.1 per cent in the UK and 51.7 per cent in Germany (Bank of Japan, 1998).

The trade union context

Union members accounted for 23 per cent of the 65 million workers employed in 1996 (31 per cent in 1980), a little higher than in the US

(15 per cent in 1996) and lower than in the UK and Germany (36 per cent in 1995). The unions are organised on a company basis and fewer working days are lost through disputes than in the US, the UK and Germany (Ministry of Labour).

Complementary institutional relations

There is good degree of cooperation between complementary organisations in the social environment and strong competition between similar businesses. Relations between political parties and government bureaucrats, business and government, business and unions, business and universities, and business and families are cooperative; however businesses in the same line are highly competitive. What conditions determine this cooperation between complementary organisations? First, each organisation recognises its mission and tries to see its role in the wider environment. Second, there are plural organisations with the same functions – not only public and private business enterprises but also public and private universities and labour federations. They tend to be flexible and competitive and in order to solicit more members, each organisation has to improve its performance. Third, the government plays an important role as a coordinator.

Fourth, the daily newspapers are centralised and powerful in Japan. There are three major and two minor national daily papers. The *Asahi Shinbun* has the second highest circulation, selling 8.4 million copies a day. The quality of this paper is high compared with the *Daily Mirror* or the *Los Angeles Times* (we would not compare it with the *New York Times* or *The Times* because the readerships are quite different – the *Asahi Shinbun* has a mass circulation). These national newspapers are powerful and function as a means of social feedback to ensure the legitimacy of business organisations' operations. When pollution became a problem the newspapers reported the facts every day and criticised the activities of business, and as a result pollution control in Japan became one of the strictest in the world.

The economic context

The economic growth rate declined markedly during the 1990s. Between 1970 and 1980 the real growth rate was about 10 per cent, from 1980–90 it was about 5 per cent and from 1990–99 was about 2 per cent. Because of the earlier growth the average income in Japan became the highest among the major countries (Table 1.3). The Japanese economy has encountered many crises in the past but it has overcome them successfully. In 1971 the exchange rate fell from 365 yen per dollar to 300 per dollar (the 'Nixon Shock'). In 1973 the oil crisis caused the price of oil to rise about four times. Inflation followed and in 1980 the second oil crisis brought another oil price rise. Above all the value of the yen against the dollar continued to slump, and by 1997 its value was less than half of what it had been in 1985, a fall not experienced in any other country (see Table 1.3).

Yet because of their tremendous effort to curtail costs and improve quality, many manufacturing companies have maintained their position in the world market. Japanese VCRs, colour TVs, fax machines, cameras and semiconductors dominate the world markets, and technology-intensive goods such as machine tools, robots, ships and steel are highly competitive. All these products are the result of excellent original research and a good interface between the development, marketing and production departments, as well as the work of highly skilled employees. Because of its competitive strength, Japan has maintained a huge positive trade balance of more than $100 billion for many years (Table 1.3).

There are many weak sectors, however. Generally speaking the productivity of non-exporting industries is low, as is that of the service sector. The upside of this, however, is that in Japan it is still possible to get good services, but at high prices, and the service sector plays a major part in maintaining and generating employment. The productivity of the agricultural sector is also low because it is government-protected. The dominant Liberal Democratic Party maintains its power base through a series of coalitions and alliances, an important one of which is with agricultural interests: hence that sector is remarkably protected and relatively unproductive.

The price of land is high – $3000 per square meter being the average price in urban areas. This is in part because of the difficulty of switching land use from agriculture to other purposes under the regulatory regime; and in part because of the important role that real estate has played as equity against loans. One positive consequence of the high cost of land and the change-of-use problems is that many small workshops have survived, producing inputs for larger firms. Japanese supply chains are long and quite complex.

Residential space is also used economically: Japanese people do not live in large suburban homes on quarter-acre plots, as do many Australians or Americans. Their dwellings are small, and because of this space constraint Japanese firms produce compact domestic consumer goods. Hence the great expertise that many Japanese firms such as

Sony have achieved in miniaturisation owes much to the requirements of the domestic market.

The savings rate of Japanese households is very high. Household saving as a percentage of disposable income has been about 14 per cent since the late 1980s, compared with 5.4 per cent in the US and 10 per cent in the UK (Bank of Japan 1998). There are several reasons for the high savings rate. One is that employees tend to invest part of their large summer and winter bonus payments, which account for about 40 per cent their salary. Another is the absence of a welfare state: because people cannot rely on the state to look after them in retirement or during times of hardship they have to save for the purpose. A third reason is that there has been a high degree of trust in the state and a very high degree of political stability in the postwar era.

Household savings are mostly deposited in banks, which in turn lend the money to industries, both directly and indirectly through securities. Consequently banks play an important role in monitoring the activities of manufacturing companies. Typically the banks have a longer-term view than do shareholders: their interest is in companies growing and hence continuing to demand loans. However they are non-interventionist in that they do not intervene in management unless companies are obviously suffering from a serious management deficit. The monitoring power of banks over manufacturing companies, particularly in the case of the 'main banks', has affected corporations' long-term planning. There is a similar situation in Germany, however the power of the major German banks is much stronger because they hold the majority of proxy votes and attach directors (*aufsichtrat*) to the companies

The research context

The number and quality of university graduates are important factors in the the Japanese research system. Of the 750 000 university and college students who graduate each year, about 153 000 are science and engineering graduates, the majority of whom go into manufacturing. In addition there are about 50 000 Master or Doctor of science graduates. These figures are comparable to those in the US, but much higher than in the UK and Germany. While it is frequently said that Japanese universities are merely playgrounds for young people and require little effort from them, this does not apply to natural science students, who have to work very hard.

Research funding is a little less than in the US but four times higher than in the UK, Germany and France (Table 1.4). Research activity on this scale has resulted in a large number of patents applications. In 1995 Japan had the largest number of patent applications in the world and Japan's share of patents in the US was also very large, as Table 1.4 shows. Hence it is a fallacy to say that Japanese companies are only imitators. Japan is exporting more patents than it imports. For instance the technology trade produced payments of $4.1 billion and receipts of $5.9 billion in 1995. However, this is not to say that no problems exist in the Japanese research system. There is less cooperation between universities and industry than in other countries and Japanese universities receive little funding from industry. This is one of the main reasons why original, fundamental and creative research lags behind that in the US. There have been only five Japanese Nobel Prize winners, compared with 180 Americans, 67 Britons and 61 Germans. The Japanese do not lack creativity and lead the world in applied research, where the bulk of research activity is concentrated.

Table 1.4 R&D spending, personnel and patent applications, 1995

	R&D spending (billion dollars)	*R&D personnel*
Japan		
Industry	94	384 000
Research institutes	19	41 000
Universities	19	161 000
Total	132	586 000
US	172	960 000
UK, Germany, France (approximate)	40–50	150 000
	Patent applications	*Share of patents applied in the US (%)*
Japan	389 000	21
USA	235 000	56
UK, FRG, France	90–120 000	2–5

Source: *White Paper 1998* (Tokyo Science and Technology Agency, 1998).

Whether Japanese-style employment practices – for example respect for people, lifetime employment, and promotion and wage increases according to length of service – promote or hinder creative activity is debatable. This will be studied in later chapters. One advantage of the lifetime employment system is that research can be conducted as part of a long-term plan and a considerable accumulation of knowledge is possible. A disadvantage is that it is hard to set up independent business ventures or independent research groups because under the lifetime employment system few competent researchers are willing to enter start-up organisations. Nor are the banks interested in lending high-risk capital. The venture capitalists and innovators who characterised the early days of Silicon Valley are lacking in Japan.

The business environment

Quality

The business environment is characterised by very strict and selective demand. Japanese consumers demand high quality and refinement in their purchases. High quality is also demanded by manufacturers in respect of the components and robotics used in the assembly process. American parts suppliers to Japanese manufacturing companies in the US, for example the NUMMI plant in California, have been surprised by the strict quality requirements, for instance glass for car windows has to be absolutely smooth and free of defects. In Japan, farmers insist that fertilisers be white (which is completely unnecessary for the plants). Cameras not only have to be small but most also have a sharp automatic focus, automatic exposure, automatic winding and require no sophisticated photographic ability to produce excellent photographs. Cars have to be easy to start, even in cold weather, and must develop absolutely no engine or mechanical trouble. Consumers like to have many instruments on the dashboard and so the cars must be well instrumented and have many standard features. The notion of an 'economical defective rate', as applied in American manufacturing, is not allowed. As a result of these strict criteria, high-quality robotics, car parts, electronics components and consumer goods are produced, bolstering the international competitiveness of Japanese products.

Japanese consumers are also hungry for new products, and the sale of ones that prove popular can rapidly reach saturation point. In other words the product life cycle is short, brought about by the 'me-too' attitude of consumers and producers' imitative activities. Many manufacturers rapidly enter new product fields, creating cheaper copies of innovative products, thus stimulating demand for new products.

High-quality products were also stimulated by the Deming Prize, which had a profound and extensive influence on quality improvement by causing quality control theory to be diffused rapidly. Most manufacturing companies raced to introduce a quality control system in order to enhance the quality of their products and win the award. The award was considered an honour that would boost the prestige of the company and have a considerable impact on sales. The US followed this award system with the Malcolm Baldridge National Quality Award.

Supply factors: group orientation

Supplies are mostly sourced from fellow members of *Keiretsu*, or alliance partners, and thus large assembly manufacturers easily transplant their high-level skills to these other firms in the supply chain. Toyota has its *Kyohokai*, composed of numerous component suppliers. Toshiba and other appliance manufacturers also have *kyoryokukai* or cooperative suppliers, who supply components on a long-term contractual basis. These relationships contrast with the American-style open bidding system, although this is changing (Dyer, 1995). Alliance relationships have made suppliers of Japanese components competitive worldwide, in areas such as compressors, shock absorbers, robots and electrical components.

The *Keiretsu* also have sales channels for durable goods. For example in 1998 Toyota had 309 exclusive sales companies, with about 5000 sales points and 120 000 salesmen, while Matsushita had 5000 selective sales outlets. These allied sales channels promote the sale of new products, thus accelerating diffusion. Alliances with suppliers and sales channels enable the supply chain management to practice agile distribution management through the JIT system.

Industry structure and rivalry

Strong rivalry exists among companies in oligopolistic industries such as cars, domestic appliances, semiconductors and other technology-intensive products. There are some less competitive sectors, however, such as those protected by government regulation or licences. Rice production is protected by import restrictions, thus the price of rice is 4000 yen per 10 kilograms compared with 1000 yen for Californian rice, on the rare occasions when it is imported. (Japanese consumers believe that the properties of their rice are unique and that imported rice cannot be as good.) The retail sector is protected by restrictions on large-scale retail outlets, the transportation sector by licences, and the chemical and paper industries by cartels.

There is strong competition among export-oriented industries but less competition in the service sector and non-exporting industries. Why, then, have foreign companies not taken advantage of this favourable environment and set up operations in Japan? (Foreign direct investment in Japan is only one fifteenth of Japanese FDI in other countries.) First, the strict consumer demands and alliances between industries do not provide a favourable environment for new entrants. Second, the high land prices, high taxes on fixed-asset transactions, high corporate income tax and high labour costs act as a deterrent, although Japanese companies have to bear the same costs. The question of low foreign penetration will be discussed further in Chapter 6.

SUMMARY

The growth rate of the Japanese economy has slowed since 1990. The brunt of this has been borne by the many financial institutions that have been handicapped since 1997 by the necessity to write off bad debts. However most manufacturing companies are strong and healthy and enjoy worldwide competitive power, as evidenced by Japan's huge positive trade balance and the strong yen. The management system adopted by Japanese manufacturing companies has been highly successful, so much so successful US firms such as Hewlett Packard and 3M have adopted similar practices.

There are five main features of the Japanese management system. First, there is a long-term orientation and a global vision rather than an emphasis on short-term profits. Second, there is a high degree of collaboration between the members of alliances, and business groups, but strong competition between the various groups. In other countries, companies are more independent, particularly in terms of suppliers and sales channels. Third, organisations are organic in structure, workers are multiskilled and there is extensive interface between departments and shared core competencies (the other model is mechanistic organisation, where job demarcation is strict). Fourth, there is respect for people, and the lifetime employment and status-ladder system help to create a learning organisation (the mechanistic model, by contrast, holds that the market should provide the required human resources – people can be bought from outside and will move freely from one organisation to another). Finally decisions are arrived at by consensus. This enables employees to share their knowledge, which also contributes to the learning organisation. (By contrast the normative US model presented in management textbooks is premised on decisions being taken by individuals in authority.)

The emergent trends in Japanese management can be seen as a shift towards an organisation modelled on a 'new *gemeinschaft*', that is, premised on a new community of interest or a new type of 'stakeholder capitalism' in which several interests are vested and represented. The new model stakeholder capitalism is based on a number of features.

First, the balancing of multiple goals. A balance between ROE, long-term growth and employee welfare is sought but the welfare of employees (or 'respect for people') continues to be important because, in the high technology environment, the knowledge held by employees is of utmost importance. Employees are not seen as interchangeable resources.

Second, vertical alliances or *keiretsu*. These have traditionally been an important feature of Japanese strategy. Yet the formation of horizontal alliances with competitors has become increasingly popular. For example Hitachi is cooperating with NEC in the development of new semiconductors. The reasons for the increase in horizontal alliances include the rising development costs for high technology products and the slow growth of the economy. Economies of scale are also important. Alliances expand the boundaries of the organisation.

Third, decentralisation. Large corporations are shifting towards a moderate decentralisation of authority by creating 'companies within the company'. Hitachi has created several such companies, enabling it to concentrate its resources on growth areas. Generalist jobs are starting to give way to specialist career paths. The development of higher technology products and processes is enabling employees to exercise greater individual independence. This does not represent a return to the strict division of labour, nor is it a 'pay for the job' system. The status ladder system continues to offer opportunities for wage increases and promotion.

Fourth, a shift from lifetime employment to more flexible employment. Specialised career paths make this shift possible. At about 50 years of age, employees are consultated about the future courses of action available to them: they may choose to move to another company, or to stay with their present company until retirement age. This system is based on respect for people and is quite different from the hire-and-fire system. Remuneration is changing from equal treatment for all in the internal labour market, with small differentials between the steps in the status system, towards one based on larger differentials that more properly reflect skills and performance.

Finally, the emergence of new strategic thinking. Decision-making by consensus is not losing sway but the ability to present creative new ideas or 'strategy construction' has become important in appraisal systems.

New models have been developed to explain the Japanese management system. These models offer explanations from the viewpoint of cultural determinism, technological determinism (or the convergence model) or strategic choice. In this book we follow the strategic choice model but give consideration to the other two models.

There are significant criticisms of the Japanese system. First, that companies are too production-oriented and employees are too company-oriented, sacrificing their home life. Moreover companies neglect shareholder value and put too much emphasis on growth and employee value; thus return on net worth is extremely low. Second, that the government exerts a strong influence on companies and politicians are connected with certain industries, hence there is no real competition, and entry barriers are erected against foreign companies. Third, that Japanese companies are mostly followers rather than innovative or niche seeking. Unlike in more organic organisations (such as the sociotechnical experiments in Scandinavia and the Netherlands) formal authority is centralised. Sometimes this is seized on as an example of Japanese authoritarianism, as is the fact that there is a large power distance in informal human relations (Clegg *et al.*, 1990). From this perspective respect for people is seen as superficial and much more importance is attached to the fact that hard work and long hours are enforced, and lifetime employment is seen as being supported by a number of temporary employees and discrimination among employees. According to this interpretation, because employees are insufficiently empowered to make bold or radical proposals decision-making is incremental and imitative.

The environment in which Japanese firms operate has the following features. First, there is a large supply of university graduates, particularly engineering graduates. Second, working hard is considered to be a desirable social value. Third, there is robust and informed commentary by newspapers on social and business affairs. Fourth competent government bureaucrats guide industry and, it is claimed, protect industry from foreign competition. Fifth, the markets are large and high-income consumers boost demand, and the high savings rate means there is an abundant supply of funds. Sixth, the number of scientific researchers is about the same as in the US but almost four times higher than the UK, Germany and France. However, basic research in the universities is not always outstanding. Seventh, the strict demand for quality by consumers and industrial buyers, together with openness to innovation, have helped to elevate the quality of technology-intensive products. Eighth, vertical strategic alliances enable swift technology transfer from core companies to suppliers and the realisation of agile management practices such as JIT or supply-chain management. Finally, export industries are intensely competitive but the service industry and non-exporting industries are less competitive, so their prices tend to be higher than the world standards. This environment provides favourable conditions for Japanese technology-intensive industries but not necessarily for foreign entrants.

APPENDIX 1.1: Sources of data

Table 1.6 Breakdown of the 205 manufacturing companies in our sample

Breakdown of companies by industry	*Number of companies*	*Companies analysed in greater detail*	*Number of companies that responded to the mail survey*
Mining and construction	4	0	3
Food and fisheries	12	11	6
Textiles	16	8	7
Paper	4	4	3
Chemicals and drugs	25	15	12
Petroleum	9	5	2
Rubber	7	2	4
Glass and cement	6	5	3
Iron, steel and nonferrous metals	12	10	9
Machinery	22	10	9
Electrical appliances and precision machinery	50	18	20
Transportation equipment	30	11	13
Miscellaneous manufacturing	8	0	6
Total	205	99	97

Note: The mail survey was carried out in November 1995; questionnaires were sent to 205 companies, 97 of which responded.
Sources: Japan Development Bank, *Japan Data Handbook* (1995); Toyo Keizai, *Quarterly Report of Companies*, 2nd and 4th edns (1995); Nikkei, *Company Report*, 1st edn (1995); Toyo Keizai, *Handbook of Foreign Direct Investment Companies* (1995); Nikkei, *Company Report*, 2nd edn (1993); Mizuki Sogo Kenkyujo, *Handbook of Corporate Organizational Structure* (1993); Toyo Keizai, *Quarterly Report of Directors* (1993).

2 Corporate Governance and Top Management

ISSUES OF CORPORATE GOVERNANCE

The meaning of corporate governance

Corporate governance refers to the means by which stakeholder control is exercised over corporations (Charkham, 1994; Clarke and Clegg, 1998; Takahashi, 1995). In the past, analysis tended to focus on the separation of management from ownership (Berle and Means, 1932) and assumed that property rights were the most important source of formal power, and hence that the key stakeholders were the owners. Often the focus would be on the ways in which control passed from the owners to the stewards – the managers – as the day-to-day custodians of capital. It was their job to see that the enterprise was efficiently and effectively controlled in the interests of the owners. (Often, some of these owners were senior management, with stock options that tend to blur boundaries and align the interests of managers and owners.) In Japanese enterprises the stakeholders are not only shareholders but also banks, employees and the unions (Figure 2.1). The current tendency in Japan, however, is to place more importance on shareholder value.

Stakeholders commit resources that are important for the survival of the corporation on a long-term basis. Of the stakeholders, it is the stockholders who have *de jure* legal power to control the corporation. (On resource-based power theory, see Pfeffer, 1981; on stakeholders, see Clarke and Clegg, 1998; Charkham, 1994) Thus for the purposes of analysing current Japanese realities, we exclude consumers, suppliers, dealers and the government as stakeholders, although to some extent they all have a power stake. We refer to these as 'interest groups' with access to a specific source of power only. For example individual consumers have resources to exchange, but unless they are organised in some way they do not create power through dependency, nor do they have legal power. (For a broader view, see Clarke and Clegg, 1998, ch. 6.)

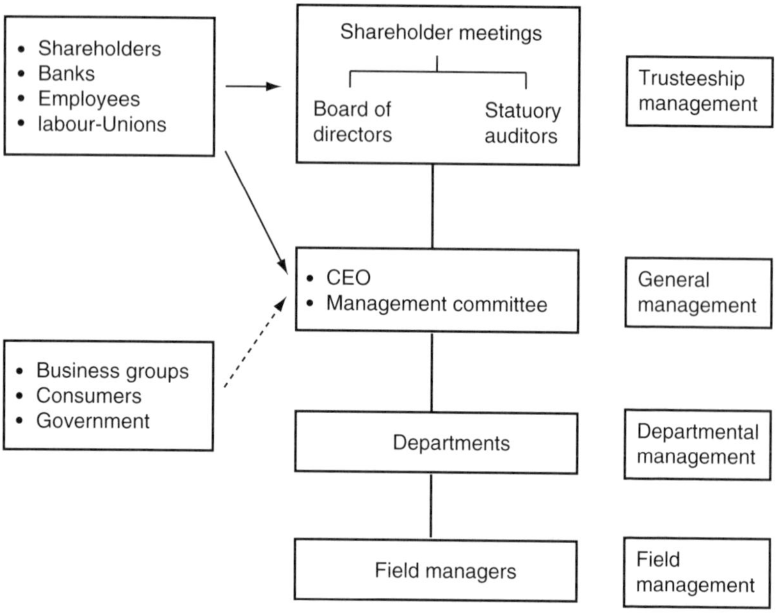

Notes:
1. Unbroken line denotes the power of stakeholders; dotted line denotes the influence of interest groups.
2. Governance means to control the structure and behaviour of the trusteeship management and the general management, and to influence policy, organisational structure and key personnel selection.

Figure 2.1 Structure of Japanese corporate governance

In this chapter we focus on the structure and process of controlling corporate governance, that is, how shareholders' meetings are controlled, the role played by the statutory auditor, how the general managers are selected and how their behaviour is monitored. The classical categorisation in the literature is that by Holden *et al.* (1941), who classified control into four levels: trusteeship management, general management, departmental management and field management. Of these, we shall examine the structure and behaviour of the trusteeship management and the general management.

Reasons for the increased concern about corporate governance

Corporate governance began to attract the attention of the wider Japanese public for a number of reasons. The unethical behaviour of top management became public knowledge as a result of a number of scandals during the 1990s, when an increase in the number of foreign shareholders led to more surveillance being exercised over share prices and governance issues than had been customary among Japanese shareholders in the past. Finally, under conditions of slow economic growth, stock prices became much more significant indicators of the health of companies.

One of the most significant examples of unethical behaviour is that of Yamaichi Securities. Yamaichi Securities, one of the largest investment companies in Japan, with 7694 employees, was bankrupted in 1998 because of its failed investment strategy. The top management had hid earlier failures and tried to recover the losses, but unsuccessfully, and eventually the losses added up to billions of dollars. These losses were caused by failure sufficiently to disclose information, a practice encouraged by past experience – Yamaichi Securities had been on the verge of bankruptcy on a previous occasion but the Ministry of Finance (MoF) had rescued it by securing a large amount of emergency finance from the Bank of Japan. This had encouraged the top management to think that, in order to protect the public, the MoF would always rescue it from errors of judgement.

Despite the ethical issues that led to their bankruptcy about 90 per cent of Yamaichi employees had found new jobs by 1999 because Yamaichi had enjoyed a reputation for competence, despite its malfeasance. Many other financial institutions had incurred bad debts because of their strong orientation towards growth. It was always assumed that asset values would grow to cover the exposure, but in the recession of the 1990s this was not the case. Domestic property portfolios shrank in value while overseas assets were ruined by the collapse of the East Asian economies after 1997. While in the past they had been able not to disclose information and could safely assume that the government would rescue them in the event of crisis, this no longer applied after the disgrace of Yamaichi.

Combined with the disgrace of respected companies such as Yamaichi, the Japanese market was becoming increasingly open. The percentage of shareholdings by foreign investors is now more than 10 per cent of the average of all companies listed on the Tokyo Stock Exchange, and in many internationally known companies such as Sony, Canon, Matsushita and Toyota that ratio has increased to more than 30 per cent. Foreign investment trust companies and pension funds require more information than do national companies. Additionally, under slow economic growth stock prices fluctuate more. In some cases stock prices have fallen from 1000 yen to 50 yen (which has been the case with many general construction companies). In such circumstances, as elsewhere, public trust in the company diminishes and in consequence the activities of the company become very constrained.

THREE MODELS OF CORPORATE GOVERNANCE

Three typical models of corporate governances are shown in Table 2.1 (see also Charkham, 1994; Takahashi, 1995; Groenewegen, 1997; Yoshimori, 1997; Clarke and Clegg, 1998). The Japanese model can be typified as an 'our company' or *gemeinschaft* model (Tönnies, 1987), where the interests of employees are give first priority and shareholders wield little power. Thus trusteeship management is weak, in marked contrast to the normative US model presented in management textbooks, where management control is strong and the interests of stockholders are represented at this level of control. Banks have more power to monitor general management practices, sometime sending one of their directors or managers to the company as a full-time board member. The company aims for long-term growth at a low profit rate, because banks are interested in expanding the wholesale financing market.

Unions have the legal power to negotiate, but since most tend to be organised on a company basis they are not so much confrontational as cooperative. These stakeholders encourage increased employment and promotion through the growth of the company. About 16 per cent of directors have experience as full-time union leaders, so top management is usually sympathetic to the needs and requests of the union.

The Japanese model can be contrasted with the other dominant models. The normative US model can be characterised as a 'your company' model, meaning that the company belongs to the shareholders, based on property rights, or as a *gesellschaft* model, where the participants commit themselves only partially, practice fair exchange and are prepared to give and take. Institutional owners such as pension or mutual funds are interested in increasing profits and share prices, and they collude to select the directors, the majority of whom are non-executive. In order for stock to be listed on the New York Stock Exchange all members of the audit committee have to be non-executive members (American Bar Association, 1994). The most important goal of the company is to increase its profits, particularly short-term profits, which will increase its share price.

Germany has a joint decision or codetermination model. The board of directors (the *Aufsichtrat*) is composed of 50 per cent stockholders and 50 per cent employees and union representatives. Although there are many large limited companies (GMBH), the joint stock company (*Aktien Gesellschaft*, AG) is the major form of large corporation, so we shall study only AGs, of which there are about 1600. The banks hold not only the stocks of the corporation but also voting rights by proxy from stockholders. Hence the banks can dominate both the general meetings and the board, many members of which are from the banks. Employees and union representatives occupy half of the board seats, but typically they do not have sufficient knowledge about, and information on, the AG's management. Even the bank representatives rarely have enough time and information but it tends to be they who give advice to the general management (*Vorstand*). The *Aufsichtrat* also has the right to select the members of the *Vorstand*. The general management members, who are the locus of managerial expertise, the *Vorstand*, are separate from the members of the *Aufsichtrat*. Since the banks are the most influential stockholders and general management has the strongest decision-making power, companies tend to emphasise long-term growth, similar to Japanese corporations. (On the German system see Charkham, 1994; Yoshimori, 1995b.)

The effects of governance structure

There are merits and demerits to each system. The merits of the Japanese system are that long-term growth is emphasised and innovations are carried out on a continuing basis. Large amounts are invested in research and development and the practice of showing respect for people means that priority is placed on employee benefit, as shown in Figure 2.2. The main demerit is that decisions about restructuring tend to be slow, particularly decisions to quit or divest, because of the community organisation concept.

The lifetime employment system promotes the accumulation of knowledge and thus helps to bring about a learning organisation. The risks attached to the Japanese model are that that the company uses a low profit hurdle rate for evaluating investment, with too much emphasis being placed on the growth of sales and job opportunities, resulting in overcapacity in respect of facilities and personnel and a low return on investment. The return on the net worth of Japanese corporations is low, on average 4.9 per cent compared with 23.8 per cent

Table 2.1 Three models of corporate governance

Basic concepts	*Japan* *'Our company'* *(Gemeinschaft)*	*US* *'Your company'* *(Gesellschaft)*	*Germany* *'Co-determination'*
Ownership dispersal	High	High	Low
Major owners	Financial institutions, corporations, individuals	Individuals, pension funds, mutual funds	Banks, individuals
Banks	Major supplier of fund (changing) Long-term vision Monitoring		Important supplier of funds, have proxy votes and long-term vision
Unions	Company-wide unions, cooperative	Industrial unions, confrontational	Industrial unions, codetermination
General meetings	Perfunctory	Reports and discussions	Perfunctory
Board of directors	99% are executive directors	Majority are non-executive directors relatively powerful,	Half are shareholders' representatives, half are employees and union representatives. Typified by lack of information and knowledge
Statutory audit	Joint auditing with CPA	None	None
General management	Decision-making by Management committee of about 10 members	3–5 top managers (CEO and senior VPs)	Decision-making by the members of the *Vorstand* (up to 10 members)

Table 2.1 *Continued*

Basic concepts	*Japan* *'Our company'* *(Gemeinschaft)*	*US* *'Your company'* *(Gesellschaft)*	*Germany* *'Co-determination'*
Educational qualifications	Many science graduates	Many finance and law graduates	Many science graduates
Remuneration of executives	$600–300 thousand (14–17 times the average pay)	CEOs receive $2.2 million plus stock options (100–150 times the average pay)	As for Japan
Financing	Debt	Equity	Debt
Major goals	Long-term growth	Short-term profits	Long-term growth
Time horizon	Long	Short	Long
Main goal	Job security	Dividends	Jobs and dividends
Risk presented to the company by the governance structure	Stock prices register resistance to the present governance structure	Staff morale lowered by restructuring and high management income	Conservative decision making

Note: Clarke and Clegg (1998) call the three models collective, stockholder and stakeholder capitalism.

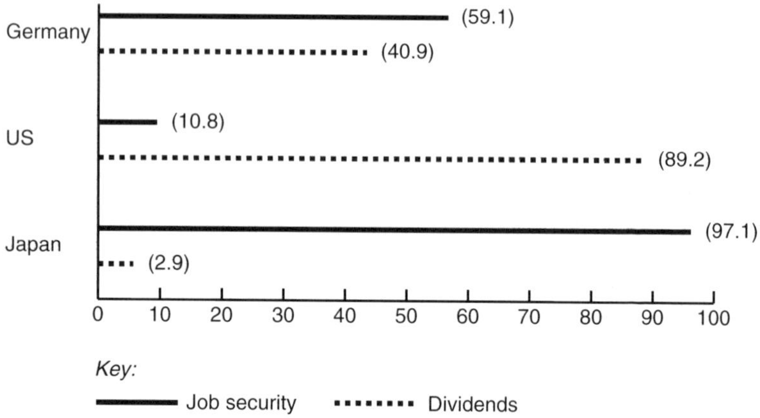

Notes: Responses to the question: Suppose a CEO must choose either to maintain dividends or to lay off a number of employees. In your country which of these alternatives would be chosen?

Number of sample companies: Germany 105, US 83, Japan 68. Mailed questionnaires were distributed in 1990–92 to graduates of INSEAD, France. The response rates were 22–26 per cent.

Source: Yoshimori (1995).

Figure 2.2 Job security or dividends (per cent)?

in the US (see Chapter 1). Because of this some companies meet resistance from the stock market or experience a sharp decline in their share price, both of which hamper the activities of the corporation. The disclosure of corporate information is restricted because of the governance structure. Sometimes, when adverse information then leaks out, this can cause a sharp decline in the stock price. In such circumstances even the main bank cannot help the company. Many financial institutions and construction companies experienced such a slump in the late 1990s, for example the stock price of Fujita, a major construction company, fell from about 600 yen in 1993 to 50 yen in 1998.

The merits of the US system is that short-term efficiency and return on investment are high, as is the resource mobility of capital, such that restructuring occurs rapidly. The disadvantages are that long-term investment in facilities and human resources are not emphasised and there is a marked difference between the remuneration received by those at the top and that paid to employees at lower levels. Those who are in strategic positions at opportune moments are more than amply rewarded. Cutting employee costs and increasing the flow of resources into the pockets of top management is a feature of this system.

The advantages and disadvantages of the German system are similar to those of the Japanese system. Companies can adopt long-term goals because banks provide long-term funds and monitor the firm with an eye to the future, and because of codetermination companies respect both employees and shareholders. However decision-making tends to be slow because of the joint decision-making system.

There is some evidence of a degree of convergence between the three systems, but strong national characteristics remain. For instance Japanese corporations are increasingly obliged to place more emphasis on shareholder value, while American corporations have discovered the importance of knowledge management and are placing increasing importance on the training of employees (for example the six-sigma movement and the training system at GE: see Groenewegen, 1997).

CONTEMPORARY TRENDS

The changes in the governance structure of large Japanese corporations are summarised in Table 2.2. The basic concept can be called the 'new *gemeinschaft*'. The power of stakeholders is changing as foreign shareholders increase in number and thus importance, while the monitoring power of banks is in decline. The emphasis on employee welfare is less strong but it has by no means been abandoned; rather the knowledge that employees can contribute is seen as the most important factor in the formation of unique core competencies. Companies are learning to balance multiple goals while maintaining respect for their people.

The average number of board directors has decreased but as few non-executive directors are appointed, boards remain insulated from pressures emanating from shareholders. General management is not changing. The management committee, which meets once a week, continues to make all strategic decisions. The title of many directors has been changed to 'corporate officer', but this is only a nominal change.

The goal of firms has changed from an emphasis on growth to the balanced achievement of multiple goals, with priority being placed on people whose knowledge constitutes a core competence. In this sense

Table 2.2 New trends in corporate governance in Japan (the 'new *Gemeinschaft*', or balancing multiple goals)

Stakeholder changes:

- Increase in the number of foreign shareholders
- Decline in the monitoring power of banks
- Increased importance placed on employee knowledge as the source of core competencies

Board of directors:

- Decrease in the number of board members to enable more effective decision-making
- Few non-executive directors, the board is not controlled by shareholders

General management:

- Group decision-making by management committee (unchanged)
- Nominal change in title from 'director' to 'corporate officer'

Goals:

- Balancing of multiple goals while maintaining respect for people and the importance of their knowledge in competence formation

the new trend is called the 'new *gesellschaft*'. A detailed analysis of these changes is provided below.

DISTRIBUTION OF SHAREHOLDERS AND BUSINESS GROUPS

Cross-holding of stocks: past and recent trends

The distribution of shareholders in major Japanese US and Germany companies is shown in Table 2.3. As can be seen, in Japan the main shareholders are banks and other corporations. The holding of stocks by individuals fell from 33 per cent in 1976 to 24 per cent in 1998, perhaps because families prefer to deposit their savings in bank or post office accounts rather than turn to the more speculative stock market. Banks and corporations engage in cross-holding or interlocking in order to increase trade or create a tie between the corporation and the main bank, a well as its suppliers and dealers. In other words, interlocking forms business groups and strengthens strategic alliances. It also protects against hostile takeover, which has become more important as stockholding by non-Japanese interests increases. Stock held for the purposes of interlocking is rarely sold, and unlike pension

Table 2.3 Distribution of shareholders (per cent)

	Japan (1998)	*US (1994)*	*Germany (1994)*
Financial institutions	40	46	29
Banks	22	–	14
Insurance companies	17	5	7
Pension funds	1	26	
Mutual funds	3	11	8
Other	1	4	
Non-financial enterprises	24	–	39
Public authorities	1	–	4
Individuals	24	49	17
Foreign corporations and individuals	10	5	12

Figures do not add up to the total, due to rounding.
Sources: Tokyo Stock Exchange; Federal Reserve System; Deutsche Bundesbank; Clarke and Clegg (1998), p. 314; Charkham (1994), p. 105.

funds in the US, these friendly stockholders rarely exert their shareholder power. The antimonopoly law prevents the banks, even the main banks, from holding more than 5 per cent of shares in companies, so the banks seek to increase the size of their loans to manufacturing companies rather than to control them. Monitoring by the main banks is restricted to ensuring that these large loans are used prudently. The banks have a long-term interest in helping companies to grow as this results in an increase in the demand for loans.

In general, given the high percentage of friendly shareholders, shareholders have little power. Stockholders' meetings, boards of directors and statutory auditors do not operate as controlling bodies, as will be examined later. Corporations place most emphasis on the long-term growth of sales, capital investment with low profit rate hurdles and benefits for employees. It is difficult to acquire other companies through hostile takeover. The existence of a small percentage of free stock at any time tends to intensify fluctuations in stock prices.

Cross-shareholding between friendly manufacturing companies and between manufacturing companies and financial institutions is extensive, as can be seen in Table 2.3. Such cross-shareholding is to some extent decreasing, but only slightly because it serves to protect firms from hostile takeover. The reasons why companies might sell the stocks of other companies are to realise non-operating profit, using the difference between the market price and the book value, and to improve the rate of return on assets.

The monitoring power of banks is declining because banks are suffering from bad debts and no longer have the same capacity to supply corporate groups with funds. However in the future the banks will probably become major suppliers of funds through the purchase of bonds or direct financing because the high domestic savings rate is continuing and most savings eventually end up in banks.

The zaibatsu groups: past and present trends

The power of the *zaibatsu* groups tends to be overemphasised, although before the Second World War they did dominate important sectors of the economy. In 1941 four groups – Mitsui, Mitsubishi, Sumitomo and Yasuda – accounted for about 32 per cent of the output of heavy industry and controlled 50 per cent of the assets of financial institutions. They were dissolved after the war as apart of the occupying power's democratisation of institutions that had been tainted by wartime militarism. Now there are six 'headless' *zaibatsu* groups – Furukawa (or the Daiichi Kangin group) and Sanwa having joined the four original groups. The groups are referred to as headless because they are not subject to family control, as in the prewar era. The two most recent *zaibatsu* have never been controlled by families. The key features of the contemporary version of *zaibatsu* are as follows:

- *Interlocking shareholdings*. About 20 per cent of member company shares are held by the group companies (Shimotani, 1993), or more-than 35 per cent in the case of the Mitsubishi group. The interlocking of stocks is not conducted with the intention of controlling other companies but to stimulate cooperation between group members, and it has the added benefet of protecting these companies from unfriendly takeover.
- *Regular strategic meetings*. Each group holds a presidents' meeting and a senior managers' meeting once a month. These meetings are used to exchange information, and not for a family or bank to control the activities of member companies. Unlike the prewar *zaibatsu* and the Korean and Philippine *zaibatsu*-type groups, there is no central family controlling the group.
- *The main bank*. The main bank plays an important role in providing capital to the member companies. This bank monitors the opera-

tions of the group companies and cooperates with other financial institutions in the group, such as insurance or trust companies.

- *Trading company*. Each group has a trading comapny to help with the exporting and other trading activities of the member companies.
- *Similar business set*. Each of the six groups has a similar set of businesses operating in the following sectors: financial, trading, construction, food, paper, chemicals, cement, iron and nonferrous metal, machinery, electrical machinery, transportation equipment, real estate and warehousing.
- *Mutual reciprocity*. If any company in the group runs into difficulties it is helped by the other companies. For example the Sumitomo Bank helped Mazda by appointing a president and supplying funds. The Sumitomo Bank also helped Asahi Brewery to recover its market share from a low of 9 per cent in 1985 to 25 per cent in1990 and 35 per cent in 1998. In addition the bank sent three consecutive presidents to Asahi, the third of whom, Tsutomu Murai, successfully introduced a number of management changes (Kono and Clegg, 1998). Representatives of the member companies in the Sumitomo group are usually encouraged to drink Asahi beer at their meetings.
- *Large combined sales*. In 1989 the total sales of the six groups, including subsidiaries but excluding financial institutions, accounted for about 20 per cent of the sales of all corporations in Japan (Shimotani, 1993). Before the war the percentage was even larger.

Benefits of zaibatsu *membership*

Mutual help and economies of scale. The member companies engage in reciprocal trade in order to stabilise or increase sales. Companies receive an ample supply of funds from the banks and can use the trading companies to help with their export activities. The latter were very important before the war, when bank funds were not abundant, as is presently the case in Korea and the Philippines. However these days the extent and merits of reciprocal trade are decreasing. For example Toshiba borrows money from banks other than those in the Mitsui group and exports goods without the help of the Mitsui Trading Company. Trade within the groups in 1989 (excluding financial institutions) was 12.2 per cent for three of the *zaibatsu* and 5.4 per cent for the other three (Fair Trade Commission, 1992).

One-set diversification. As noted above, each group comprises a set of key industries. The degree of diversification this entails requires a large amount of investment, which is made possible by belonging to a group. The one-set principle is different from a niche strategy and results in the duplication of products and hence intensified competition, the basis of worldwide competitive power.

Exchange of information and mutual monitoring. Information on political situations and industrial trends is exchanged during the monthly meetings mentioned above, and sometimes disciplinary action is carried out. For example when misbehaviour on the part of the president of a department store became apparent a non-executive director of Mitsukoshi attempted to have him removed from the presidency during the board of directors meeting. This was a very difficult task because only the president had the authority to determine the agenda, almost all of the directors had been promoted by him and he had carefully lobbied the board members beforehand. Voting was eventually carried out under 'other business'.

Some non-*zaibatsu* companies, for instance Matsushita, Sony, Honda, Toyota and Nippon Steel, have many vertically integrated subsidiary companies. These groups will be discussed in Chapter 5.

Contemporary trends

Inter-*zaibatsu* mergers of large banks are underway between the Mitsui and Sumitomo groups and between the Fuyo and Daiichi Kangin groups. The extent to which these mergers will affect corporate behaviour is not yet clear, but trade between the groups is likely to increase. New banks with large capital resources are likely to enjoy increased monitoring power if debt financing continues.

GENERAL MEETING OF SHAREHOLDERS AND THE TOP MANAGEMENT STRUCTURE

General meeting of shareholders

The general meetings of Japanese corporations are mostly perfunctory. They usually last no more than 30 minutes because most of the shareholders submit their proxy to the management and hence the issues proposed by the top management are quickly agreed. The shareholders of mutual holding groups tend not to intervene in management and other institutional holders tend not to exert control over other companies. Unlike in the US, there are no powerful pension funds, and unlike in Germany, banks do not hold a proxy vote. In the past professional troublemakers associated with *Yakuza* gangsters made money by disrupting shareholders meetings to such an extent that the management resorted to buying their silence. This practice diminished when all companies adopted the same date for their meetings (for example 2355 companies held their shareholders' meetings on June 27 in 1997) and it became illegal to buy off such agitators.

Trusteeship management

There are four levels of management: trusteeship management, general management, departmental management and field management (Holden *et al.*, 1941). The first two levels comprise the top management, which is expected to perform several functions: to define the value system and the goals of the organisation, taking into consideration the demands of shareholders and other stakeholders; to define the relationship between the organisation and the environment, and to make decisions on product-market strategy; and to determine the organisational structure and select key personnel (Katz and Kahn, 1966). Thus strategy, structure and performance are greatly influenced by top management.

The strategic function not only requires key expertise but also influences the promotion of top executives. Technology-related diversified companies require technological knowledge and in these companies technological experts obtain the power to control, while companies that are marketing-related and diversified require marketing personnel in top management (Table 2.4). For any strategy to be successful there have to be appropriate capabilities in top management and a good fit between the strategy and the technical skill of top executives.

Board of directors: past and recent trends

The membership of boards of directors of large corporations has changed drastically in recent years with the number falling from an average of just over 26 (Table 2.4) to about 10 and additional outside members being introduced. Before this change, at the trusteeship level there were typically two components: the board of directors and the statutory auditor (the latter is required by Japanese company law), neither of which functioned effectively. Boards met less than once a month and took decisions only on legally required matters. For

Table 2.4 Features of top management, board of directors and auditors (averages)[1]

President:	
Age (years)	64.0 (5.2)[2]
Disciplinary background (per cent)	·
Social science	54
Natural science	42
Other	4
Tenure (years)	9.0 (4.5)[2]
Remuneration ($ per annum, estimated)	600 000
Directors (including president but excluding outside directors and auditors:)	
Age (years)	58.6 (1.9)[2]
Tenure (years)	5.5 (1.6)[2]
Degree category (per cent)	
Social science	47
Natural science	46
Other	5%
Board composition	
Average number	26.0 (7.0)[2]
Outsiders	0.37
Percentage of graduates from two universities	51
Stock ownership (per cent)	1.65
Remuneration ($ per annum, estimated)	300 000
Auditors (number):	
Full-time	2.70
Part-time	0.39
Total	3.09

Notes:
1. Averages based on 99 companies.
2. Standard deviation.
3. In 1993, 2574 directors and auditors, excluding outside directors and auditors were analysed for this table.

Source: Based on Toyo Keizai's *Yakuin Shikiho* (quarterly report on directors) (1993).

instance long-term plans were seldom submitted to boards for authorisation (only 22 per cent were submitted in 1989 – survey carried out by Kono).

There were several reasons why the boards did not serve as the actual governing body, including the perfunctory nature of the general meeting and the lack of interest group participation in the selection of board members. Most directors were executive directors, and there were only one or two non-executive directors on a typical board of 26 members – there was no legal requirement on this matter. In the US the majority of directors are non-executive and the New York Stock Exchange requires listed companies to have an audit committee composed solely of non-executive directors (American Bar Association, 1994) – in Japan there is no such rule. In Germany the codetermination law specifies that the board members (*Aufsichtrat*) have to be separate from general management (*Vorstand*) and composed of shareholders and representatives of the employees and unions.

The meaning of 'director' in Japan is different from that in the US or Germany and has higher prestige than just 'head of department'. Companies reserve this precious title for successful inside managers rather than give it to outside persons. Almost all large corporations have management committees to carry out strategic decisions that reflect the interests of employees, banks and shareholders. The dysfunctional aspects of this are that there is no one to check the power of the president (CEO) and the interests of shareholders tend to be neglected.

Recently a number of prominent companies have led the way in slimming the ranks of their corporate boards. For instance Sony had 38 directors in 1997 but this number has now been reduced to 10, of whom three are non-executive. At the time of the changes Sony sent letters to the removed directors' families to reassure them that the change did not constitute a demotion. In the transition the title of 31 directors was changed to 'corporate executive officer', which corresponds to 'vice president' in American corporations. Corporate executive officers have been further categorised as senior, superior or junior. Likewise in 1999 Hitachi reduced the number of its directors from 30 to 14 and created 21 corporate executive officers, Toshiba reduced its directors from 32 to 12, selecting 15 corporate executive officers, and Taisei Construction reduced its directors from 49 to 12 and created 37 corporate executives. In all of these cases, no outside directors were appointed. In 1999 hundreds of large corporations reduced their number of directors to about 10 in a clear case of institutional isomorphism: a structural change driven by a changing sense of what is appropriate to the culture of the time. While we anticipate that this trend will accelerate, its effects are not entirely clear. For one thing, management committees staffed by full-time executives are to remain, supported by a large number of staff at headquarters (about 1000 at Sony). These committees meet frequently and seem likely to be the real locus of strategic decision-making. The most important effects seem to be that boards will become more involved in real decision-making, more inclined to focus on increased profits and cash flow rather than just expansion, more open to a wider range of views through the involvement of non-executive directors, and more likely to disclose information. Thus the board of directors will remain a major monitoring agent but it is not likely to become a strategic decision-making body.

Statutory auditors

Corporation law requires the use of three statutory auditors. These people are selected at the general meeting, they have a three-year tenure and one third of them must come from outside the company. The auditing is done in cooperation with outside auditing companies, which send Chartered Professional Accountants to the corporation. Statutory auditors can also conduct management audits and may attend board meetings to obtain information. The statutory auditors system is unique to Japan. It developed in an ancillary capacity to help CPA auditing and protect the interests of shareholders. The auditors are not CPAs nor are they lawyers, so they are not qualified for either strictly financial or strictly legal auditing. However if they are all selected from outside they function in a similar manner to the audit committee of an American board.

Management committee

In Japan the management committee, composed of senior full-time directors, is the most important executive decision-making body. Holden *et al.* (1941) classified general management levels into four types: chief executive, management committee (council of general executives), assembly of department heads (chief executive and council of divisional executives) and board of directors. In Japan, management committees were introduced during the 1950s by most large corporations as a group decision-making body. According to Kono's (1984a) survey of long-term planning, at 76 per cent of companies the long-term plan is reviewed by the management committee and at 22 per cent it is reviewed by the departmental heads (Kono, 1989 survey).

The management committee meets once a week for a morning or afternoon. The clerical office for the management committee is usually the planning department, which arranges the topics and materials. The management committee is neither an advisory committee to the president nor an information dissemination body but a decision-making body. Management committees take the final decisions in 52 per cent of companies, the president is responsible in 35 per cent of companies and the board of directors in 22 per cent of companies (Kono, 1989 survey). The president does not force a consensus.

There are some differences between Japan and the US in terms of the nature of management committees. In the US they usually comprise only three or four people and the meetings are somewhat informal, often taking place in the CEO's office, and senior VPs may join in the discussions from time to time. In the US it is usually the CEO who takes decisions, after consulting with other VPs. In Japan the committee usually comprises about 10 people and a regular formal meeting is held once a week to reach decisions by consensus. Many of the members of management committees are not heads of department but have broader responsibilities. Information is sought from and advice given to several departments, but the management committee does not have the authority to take decisions on departmental matters. Rather its members are in charge of general management.

Recently in Japan the tendency has been to reinforce the top decision-making bodies. In addition to the ordinary management committee, many companies have set up a senior management committee. This is composed of five to eight members, including the chairman, the president and a few senior executive directors. The senior management committee discusses only strategic problems and refers specific issues to the management committees for the final decision. In some companies strategic issues are discussed and decided by the senior management committee and operational issues are discussed and decided by the ordinary management committee.

In addition to the general management committees, in many companies departmental heads meet to exchange information, and functional committees specialising in research and development, marketing and production coordinate the activities of the various divisions. Overall the style can be characterized as one of group decision-making.

While it might be thought that group decision-making would tend to be conservative and mediocre because of 'group-think', we would argue that it in fact involves considerable risk-taking. Because of the group discussions much more information becomes generally available, hence uncertainly is reduced and committee members are more willing to take risks (Clark, 1971). Group decision-making frees individuals from full responsibility since the group always shapes the final decisions. Group discussions provide a context in which a higher value is attached to risky proposals than to conservative suggestions because the responsibility for the final decision is never one person's alone. Hence relatively cautious participants may be induced to make moderately risky proposals. The tendency to take risks is most marked when the consequences of failure are not severe. In Japanese corporations, group decision-making takes place at every level of the organisation and the overall result has been the promotion of innovation. The collection and wide dissemination of information has led to more rational and better informed decision-making, in contrast with the situation where one person is responsible for taking decisions based on less information or on personal intuition.

If the decisions of the management committees are to be effective the team members must be well informed about the issues in question before the meeting starts, either through material issued beforehand or through verbal explanations delivered to key personnel. This verbal communication is sometimes called *Nemawashi* (meaning to dig round the root of a tree), 'lobbing' or 'log rolling'. Such a process is a necessary part of group decision-making. It is not political manipulation: the discussions by the management committee may be short and look like a mere formality, but this does not mean that decision-making is a mere formality. For group decisions *Nemawashi* is necessary.

Top managers: directors and corporate executive officers

As discussed above, the title of many directors was recently changed to corporate executive officer, cutting the number of directors in large companies to about 10. The majority of directors are full-time organisation members. They include the chairman, the president and the chief executive officer, senior executive directors and executive directors in charge of several departments, and ordinary directors who head important departments. Thus to study directors is to study general management and key departmental managers. In American terms, their roles correspond to those of chairman, president, executive vice president and selected key vice presidents.

AGE OF DIRECTORS AND CORPORATE EXECUTIVE OFFICERS

One might expect that, under the length of service system, promotion would be slow and in consequence Japanese executives would be elderly. However a MITI survey in 1996 of 454 large manufacturing corporations found that the average age of directors was 58.3 years. According to our survey, in 1993 the average age of the 2574 directors of the 99 largest manufacturing corporations was 58.6 years – approximately the same as the average age of vice presidents in large US corporations.

The rapid promotion of elite managers to the post of director at a young age is rare. The standard deviation for the age of directors is small (1.9: see Table 2.4). Under the lifetime employment and internal labour market systems, overly rapid promotion results in too long a tenure as director and reduces the promotion opportunities for others, hence it is avoided.

Key skills for promotion

Top managers need to have conceptual, human relations and technical skills (Katz and Kahn, 1966). In Japan the first two are the most important, though general managers must have some specialised skill before they can be promoted to the top. Once there, however, their technical skills are of secondary importance to their conceptual and human relations skills.

The educational background of the 2574 directors of the 99 large manufacturing corporations surveyed in 1993 (Table 2.4) was as follows: social scientists, 47 per cent; engineering and natural science, 46 per cent, other disciplines 5 per cent. The remainings 2 per cent had no disciplinary allegiance or came from the outside organisation. The percentage of engineers in technology-related companies ranged from 41 per cent to 49 per cent (41 per cent for marketing and technology-related companies and 49 percent for purely technology-related companies. Top managers with a natural science background tend to be more innovative because they can understand technological change more easily.

That 51 per cent of graduates came from just two universities, they vary from company to company, is surprising and the reasons for it are not clear. One explanation might be that these are high-prestige universities, are large and thus have many graduates to choose from. Another might be that it makes sense for the most prestigious firms to recruit mainly, if not exclusively, from the most prestigious universities.

Top managers usually have particular areas of expertise (Table 2.5). On average top managers' major skills are in technology and production (35.9 per cent) or sales and marketing (32.6 per cent). Expertise in finance and administration is not high, unlike in the American system. The success of Japanese companies in manufacturing technology-intensive products may originate partly from the composition of top management in respect of expertise.

Top managers also need conceptual skills, which are gained by working in several unrelated departments, for example the marketing, production and finance departments, or the production, research and marketing departments. Managers are required to have broad experience in order to be promoted. Another important attribute in terms of promotion is experience as a union leader. About 16 per cent of top managers once served as leaders of the local union offices or the central union office, and this experience benefits mutual understanding between the top management and the union.

Remuneration

Before the war, top managers in Japanese corporations jointly received 5–10 per cent of the company's profits as a bonus. If this bonus was

Table 2.5 Areas of expertise of presidents and directors (per cent)

	Directors (1990)	*Presidents (1996)*
Technology and production	35.9	38.7
Marketing	32.6	33.1
Finance and administration	13.4	9.7
Personnel	6.7	6.3
Planning	9.8	11.3
General affairs	1.6	0.9
Total	100.0	100.0

Sources: Survey of manufacturing companies: 466 companies in 1990 and 454 companies in 1996; MITI, *Indices of Management Capabilities* (1990, 1996).

distributed just to the top 10 managers, the amount received by each was easily 100 times the pay of young employees. Today the gap has closed somewhat: the average income of directors is estimated to be about 30 000 thousand yen per year (about US$300 000), which is about six times that of the average employee. A share of the profits is still distributed to directors, but the amount is much smaller, usually less than 1 per cent.

Fringe benefits are not extensive in Japan. Companies provide a driver service but not the car itself. The cost of entertaining company guests is reimbursed rather than covered by a regular allowance. Housing is not usually provided. Stock options have been increasingly offered since companies became eligible to hold capital stock, but only to a limited extent. In the US restructuring or reducing the number of employees may increase profits and raise the value of the stock options and bonuses received by top managers, something that rarely happens in Japan. Instead companies try to maintain their employment levels and cut the bonuses of top management.

The incentives for hard work are more moral than financial. Top managers work hard because, in large part, they have a strong sense of mission that is developed through their long years of service to the company. Another is the high social prestige attached to being a company director. Extending the retirement age until 65 years is another source of moral incentive because managers want to stay in the company as long as possible, rather than retiring at what was, until relatively recently, the normal age of 60.

Stock ownership

Directors are not selected to represent stockholders, but are chosen for their particular abilities. They own few shares in the company. In 1993 the stock owned by the 2574 directors of the 99 largest manufacturing corporations accounted for only 1.65 per cent of all outstanding stock.

However some companies, particularly those with foreign equity such as Sony, now provide stock options to higher level managers to encourage them to improve shareholder value. Since the Corporation Act was changed to allow companies to hold treasury stock, managers have been able buy treasury stock or new issues of stock at the option price. The stock option and the bonus paid to managers accounts for more than half of their remuneration, thus to a large extent their salaries reflect the company's profits. It does not appear that these changes were intended to transform corporate philosophy, however. Top managers are not being required to 'cut the neck of employees to enrich [their own] pockets' (Groenewegen, 1997), and companies still seem to be trying to maintain their employment levels as much as possible.

The president

In Japan the president is usually the chief executive officer. According to our survey (Table 2.4) the average age of presidents is 64.0, a little higher than the average age of American chief executive officers. The tenure of presidents varies from company to company, but according to our survey the average is about nine years (with a standard deviation of 4.5 years), much longer than that of a director. Educationally, 54 per cent of presidents are social scientists and 42 per cent are natural scientists. The average remuneration is 60 000 thousand yen per annum (about $600 000), about 12 times the average annual pay of employees but much less (about one fourth) than the amount received by American CEOs (although almost the same as that paid by German and French companies).

The power of the president is relatively strong. He usually selects the other directors in consultation with the chairman (it is usually a he: few women achieve this rank in Japan), who is normally in charge of external relations. If the president, attends management committee meetings he usually presides over those meetings. There are no non-executive directors on the management committee, and as there are very few non-executive directors on the board of directors there is little countervailing power to that of the president. One might think that this would lead to abuse of power, but it does not.

TOP MANAGEMENT DECISION-MAKING AND LEADERSHIP STYLES

Intimate knowledge of the Japanese scene combined with a review of North American literature (Mintzberg, 1973; Steiner and Miner, 1977; Miles and Snow, 1978) suggest there are four decision-making and leadership styles among top managements (Table 2.6).

Table 2.6 Top management styles

1. Innovative and analytical	*2. Innovative and authoritarian*	*3. Conservative and analytical*	*4. Conservative and intuitive*
Top management team			
Devoted	Selfish	Theoretically consistent	Selfish
Aggressive and innovative	Aggressive and innovative	Idealistic and perfectionist	Conservative and tradition-oriented
Sensitive to new information and new ideas	Intuitively sensitive to new opportunities	Sticks to principles, theoretical	Sticks to past experiences
Flexible about ideas, capable of presenting many alternative	Intuitive in ideas, only capable of presenting few alternatives	Suboptimal and incremental decisions	Inflexible
Able to make quick decisions that they can integrate rapidly with prior policies	Tends to make quick decisions without full consideration of resources	Reluctant to make decisions until certain of information and resources	Tends to delay decisions until serious lacuna in decision-making are discovered
CEO			
Clear about setting out goals and guidelines	Tends to implement measures by himself, rather than setting out goals	Setting rates rather than goals.	Lacking in goals
Good two-way communication	One-way communication	Specialists collect information.	Little communication
Ready to listen to others' opinions	Ready to enforce his opinion, feared by others	Enough information is collected.	Allows freedom or requests obedience
Tolerant of failure	Critical of failure	Deduction system, failures are punished.	Applies no sanctions (weak leadership) or punishes deviations (authoritarian)

Table 2.6 Continued

1. Innovative and analytical	*2. Innovative and authoritarian*	*3. Conservative and analytical*	*4. Conservative and intuitive*
Examples:			
Honda, Sony, Canon	Newly established companies: Kyocera, Nintendo, Secom; bankrupt companies: Eidai, Fuji-Sash	Public organisations, old large corporations	Numerous companies bankrupted because of overly conservative decision-making

The innovative and analytical type

For presidents who exemplify the innovative and analytical type, innovation is the most important value. Typically, they encourage the collection of large amounts of information about future opportunities, keep a close watch on the business environment, are sensitive to new opportunities and generate alternative ideas based on information, ideas and data collected from both outside and inside the organisation. Decision-making depends on forecasts of the risks involved. The leadership style is typically participative – by listening to the opinions of subordinates the president strives to create a free atmosphere and is generous about failure – the emphasis is on innovation, not success.

The top management at Canon, Sony and Honda exemplify this type. Canon has engaged in comprehensive, long-range planning for many years, collecting a considerable amount of information on the environment and considering many new ideas. In this sense its decision-making style is analytical. The company spends more than 10 per cent of its sales receipts on research and development. Canon started out as a camera manufacturer but now cameras account for only 20 per cent of its total sales. Other products include copiers, printers for personal computers, word processors and related goods. Sales and profits have increased continually for 20 years. At Honda the president usually retires at about the age of 55. The sons of directors are not allowed to become employees – this is aimed at preventing nepotism and ensuring constant innovation. This type of top management produces a culture of vitality, resulting in better financial performance and better employee satisfaction.

The innovative and authoritarian type

Here the leadership style of the president is forceful and authoritative and this is reflected in the behaviour of other senior managers. The decision-making style is innovative and intuitive – ideas come before the collection of information. The president requires directors and other subordinates to obey and implement his directions as a matter of course. Subordinates are required to have a strong ability to carry out directions rather than show initiative. Newly established companies such as Kyocera (semiconductor packages), Nintendo (game machines) and Itoyokado (supermarkets) exemplify this type.

Innovative and authoritarian presidents are typically found in family-owned corporations, both old and new. Such presidents are indispensible when companies go through a drastic change. For example when Shiseido, the cosmetics company, experienced a crisis the new president, Fukuhara, used his strong leadership abilities to restructure the management. Mitsubishi Heavy Industries managed its crisis by changing its leadership from a bottom-up style to a more top-down style.

When financial performance is outstanding then the decisions of the top management will be seen as correct. However if aggressive decisions by top management seem inappropriate or erroneous when financial performance is poor, then the overall culture will deteriorate; typically the authoritarian tendencies of the president will be blamed. A number of companies have been bankrupted as a consequence of overly aggressive decision-making by dictatorial presidents, such as Fuji Sash, Kojin (textiles) and Eidai (laminated hardboard and housing).

The conservative and analytical type

Here the top management sticks to general principles, is very analytical, collects too much information and will not, in the idiom of Japan, even 'cross a safe stone bridge'. Many rules are created and they are expected to be followed. Hence it is a bureaucratic type of management, with all the predictable problems of rule boundedness. When the top management carries out a new strategy it does so incrementally and leadership is weak. This can be seen in old, large companies such as the former Japan National Railways and many other public enterprises and organisations. Examples in the private sector are Toyo Electric, Tobu Railways and Nikon.

The conservative and intuitive type

With this type the decision-making style is tradition-oriented and conservative and top management rarely initiates new ventures. Innovation occurs only when a problem arises, and leadership is very weak. For example in the large, family-owned, Ataka Trading Company top management was almost non-existent, – there was a long tradition of family rule, but the last of the presidents was more interested in collecting paintings and classic cars than in management, and as a result one crucial transaction bankrupted the firm: crude oil was sold to an oil refinining company that went bankrupt and the payment could not be collected.

In other cases, the top management wields considerable power. Decisions tend to be based on insufficient information and are characterised by conservative and inflexible reasoning. The decision-making style of Henry Ford during the later stages of his company presidency is a typical example: think of the famous statement 'we will sell any colour of car as long as it is black'.

SURVEY OF TOP MANAGEMENT BEHAVIOUR

Leadership styles and skills

The leadership styles that presidents use to arrive at a decision in the management committee are shown in Table 2.7. The data was obtained from a large-scale questionnaire survey carried out by MITI in 1996. The president-centred styles, which are similar to type 1 in Table 2.6 were fairly common but the consensus style, similar to type 2 in Table 2.6 was most common and resulted in the best performance. There were no equivalents to types 3 and 4.

The skills required of top managers can be gleaned from a survey by the *Nihon Keizai* newspaper, which conducted a series of interviews with the presidents of 75 large manufacturing companies. The details of what they considered to be important personal characteristics of presidents are shown in Table 2.8. As can be seen, vision, initiative, willingness to listen to others' opinions and impartiality were seen as important. In terms of the management types detailed in Table 2.6, it seems that types' 1 and 2 two were considered desirable. Note that health was also considered important because of the heavy burden of the duties involved in being president.

Table 2.7 Leaderships style of the president in group decision-making situations

	Per cent
The president takes the final decision, but takes into consideration the opinions of directors	17.3
The president takes the final decision, based on the majority opinion (majority rule)	13.7
The committee members participate as equal partners; concensus is arrived at under the leadership of the president	69.0

Source: MITI (1996).

Table 2.8 Skills and behaviour required of a company president[1]

	Number	*Per cent*[2]
Conceptual skills		
Broad vision, international vision	15	20
Long-range vision and flexibility	28	37
Aggressively initiative and capable of making decisions in risky situations	24	32
Hard work and hard study	6	8
Human skills		
Clear statement of goals and guidelines	20	27
Willingness to listen to others' opinions	18	24
Impartiality, unselfishness and faithfulness	28	37
Ability to use employees' capacity fully	18	24
Attractive personality	11	15
Ability to build up a team and create harmony	13	18
Ability to vitalize organization culture		
Health	24	32

Notes:
1. Opinions stated by the presidents of 75 large manufacturing companies.
2. Number of items stated, divided by the number of presidents.

Source: Series of articles on top management in *Nihon Keiai Shinbun* in 1980 and 1989.

Other models

Charismatic leadership can transform an organisation (House, 1976; Bass, 1985). Charismatic leaders are able to raise their subordinates' awareness of the importance and value of designated outcomes. They make issues clear and persuade others to transcend self-interest for the sake of the team. They motivate personnel to provide a dedicated service to the organisation, and create personal trust through sound decisions, building up self-confidence and devotion. These characteristics are similar to those found in the innovative and analytical types of top management in Table 2.6.

Collins and Porras's (1994) 'Visionary Company' is similar to type 1, but some features are added. This includes a corporate philosophy that emphasises 'and' rather than 'either/or' (for example profit and growth, not profit or growth), and a corporate culture in which the practice of trial and error is encouraged. Present success is never enough: the company readily invests resources for the future.

Factors in the formation of management behaviour

The behaviour of top management can be seen as being shaped – rather than determined – by the organisation's competitive environment, its stakeholders and other interest groups, as well as by individual attributes of the managers in question. Where competition is strong, decisions need to be innovative, while companies that enjoy a monopoly can indulge in conservative decision-making. Where the power of shareholders is strong the top management tends to emphasise short-term profits. Where the main banks have considerable power, long-term growth is emphasised. When the expectations of the business group are strong, the 'me too' strategy tends to be popular. In companies where the board of directors mostly consists of company executives the interests of shareholders tend not to be represented.

Natural science graduates tend to have a better understanding of technology. Top managers whose careers have been in production or marketing tend to be more oriented towards innovation than those whose careers have been in finance or law. Older, longer serving manager, tend to be conservative. When stock options are part of executives, remuneration they tend to be more interested in short-term profits and in boosting the stock price.

MANAGEMENT PROBLEMS

Disregarding shareholder value

The power of shareholders has traditionally been very weak in Japanese corporations because stockholding is scattered rather than concentrated, shareholders' meetings are only a formality as most shareholders proxy their vote to management, and boards of directors are mostly composed of company executives. However the percentage of foreign shareholders is increasing and these shareholders want to have a voice. During the period of slow economic growth in the 1990s stock prices slumped. Many companies were over capitalised, overstaffed and overextended due to their heavy emphasis on growth, and consequently profits declined. Some stock prices fell by 90 per cent in 1998, as was the case with many construction companies, which were particularly hard hit by the recession in construction caused by overinvestment in and overvaluation of property portfolios. The decimation of stock prices hampered the financial management of many corporations.

Good governance involves giving monitoring power to all stakeholders and interest groups, such as shareholders, banks and employees. These stakeholders should be treated equally and fairly with respect to the distribution of wealth, risk and monitoring power. While the concept that power resides only in property rights has not been characteristic of Japan in the past, awareness of the need to pay more attention to shareholder value has increased in recent years.

Insufficient disclosure, resulting in unethical behaviour

Shareholders' meetings last only 30 minutes and almost all members of the board are executive directors, which often means that failed investment strategies and manipulated financial statements do not become evident until disaster strikes. Disclosure is not a strong imperative under these circumstances. In order to conceal failure other investments are tried, often exacerbating the damage. Thus unethical behaviour is followed by further unethical behaviour.

Over reliance on groups

One argument has it that government–business alliances (Japan Inc.), cross-shareholding and the existence of business groups such as the *zaibatsu* tend to foster reliance on the group, like a ship in a convoy. In relying on the group a company may lose its focus, neglecting opportunities to carry out restructuring. The one-set principle means that each group has a similar mix of industries and thus there are many companies in the same industry.

This argument is open to discussion, as there are evident merits and demerits to the criticism. Group-oriented companies can move swiftly into growth areas, resulting in a strongly competitive environment. Those companies that survive domestic conditions thus develop world-wide competitive power. The disadvantage is that divestment tends to be slow and inefficient companies survive longer than they should.

Too many directors and too few non-executive directors

On average the top 200 manufacturing companies have 26 directors and 0.3 non-executive directors, unlike in the US, where the average number of directors is about 10, the majority of whom are outside non-executive directors, and Germany, where the *Aufsichtrat* is composed of non-executive members who are not members of the *Vorstand.* While the Japanese practice results in the neglect of shareholder value, in the US there is too much emphasis on short-term profits and in Germany outside directors have insufficient information (Yoshimori, 1995b).

However in Japan the management committee, with about 10 full-time members, is the most important decision-making body, so the problem with the board of directors is not as great as it might be. Furthermore there is the new trend to reduce the number of directors and appoint non-executive directors to represent the interests of wider stakeholders, as discussed earlier.

SUMMARY

Corporate governance refers to the means that stakeholders use to control and monitor the company. The evaluation criteria refer to how stakeholder expectations can be integrated into the strategy and operation of the company, the distribution of value, risk bearing, and how power is monitored. The stakeholders hold the key resources for the survival of the company and represent the dominant interests in the company on the board, as well as having the legal power to control it.

It is too simplistic to define control only in terms of property rights. We propose three major models of corporate governance: the 'our company', 'your company' and 'joint decision' models (Table 2.1).

In Japan, cross-shareholding for the purpose of forming a business group is popular. The six *zaibatsu* business groups account for about 20 per cent of the total sales of all corporations. In addition there are many vertical business groups, which create a larger scale of activity for the firms that are related, as they provide each other with business. However, this strategy limits agility and opportunities for diversification. It is the latter which contributes to the high degree of institutional isomorphism among Japanese business groups.

Mergers by leading banks in different *zaibatsu* groups have been common of late. For example Mitsui Bank is the result of a merger between Sakura Bank and Sumitomo Bank, and Fuji Bank (The Fuyo group) was formed by the merger of Daichi Kangin Bank and Nihon Kogyo Bank. These merged entities can be expected to change business behaviour in the groups and expand their boundaries, as well as to increase cooperation between companies within the expanded groups.

General shareholders' meetings tend to be perfunctory. Boards of directors are composed almost entirely of executive directors. Thus top management has real power and is insulated from interference by shareholders. It is this concentration of power that enables companies to look to the long range. Group decision-making by the management committees tends to encourage risk-taking. About 50 per cent of directors (equivalent to CEOs and vice presidents in the US) are science graduates and the majority have production and marketing backgrounds, rather than financial or legal experience. Such expertise allows the company aggressively to develop new products and expand into overseas markets. Typically there is relatively little differentiation (compared with the US or Australia, for instance) between the remuneration of the top management and that of the average employee: a factor of about seven, or, in the case of a CEO, 14.

There are four types of top management: innovative and analytical, innovative and authoritative, conservative and analytical, and conservative and intuitive. Successful companies seem to start off being innovative and authoritative and then shift to being innovative and analytical. The causal factors shaping these types are the governance structure (or the separation of management from shareholders), the executives' career and educational background, and the remuneration system.

Changes in governance structure are underway in Japan. The predominant style of Japanese corporate governance, with its lack of emphasis on shareholder value, became problematic as a result of the increase in foreign shareholders, the decline in stock prices during the recessionary 1990s and unethical behaviour in some high-profile companies. Negligence of shareholder value these days sometimes results in a sharp drop in share prices – which was rare in the past. The composition of boards of directors, with too many inside directors, is another problem. Many companies have started to reduce the number of directors and appoint outsiders to the board in order to reflect the expectations of shareholders, anticipating that the hurdle rate for investment will be raised, as well as going some way towards improved disclosure of information. However management committees have been functioning well in many corporations and we do not anticipate that the innovative style of successful Japanese corporations will change very much as a result of this change in the composition of boards.

3 Goals and Philosophies

CORPORATE GOVERNANCE, GOALS AND CREEDS

Top management usually create corporate goals and creeds in the early history of the corporation that capitalise on their knowledge and experience. As the corporation develops, such creeds have to integrate the values of multiple stakeholders and interest groups. In Japan the interests of banks, employees and the company president have been most strongly represented, and this representation has been translated into an emphasis on long-term growth. However since the economic no-growth period in the 1990s the power of shareholders has increased. Top management initiate and make final decisions on corporate goals but they do not so unilaterally. Changes in the value and power of stakeholders and other interest groups, as well as changing opportunities in the market, have to be taken into consideration. These changes need to be reflected in the business creed: to this end many companies have created a study team of young managers whose task is to undertake a periodic revision of the creed.

Categories of corporate goals

Various features of corporate goals are shown in Table 3.1. The basic goals comprise the core value or 'system' goals of the organisation, such as growth, profit and stability (in this book we do not differentiate between goals and policies). Long-term goals and visions are explicitly stated. Product-market strategy, organisational structure and personnel management policy follow these basic goals, and rules and standards are set within their framework (Figure 3.1). The concept of mission refers to the image of itself that a company wants to project into the wider world as depicting some aspect of its core business. The 'water supply principle' of Matsushita Electric in Figure 3.2 is one example. Missions support the basic goals and are useful in attaining them. However companies cannot survive just by pursuing their mission.

Social responsibility implies that a firm is prepared to sacrifice short-term profits in pursuit of longer-term, sustainable business. For instance a policy to prevent environmental pollution and a policy

Table 3.1 Features of corporate goals

Features	*Merits or effects*	*Problems*	*New trends in goal setting*
Long time horizon	• Investment in training and R&D	• Neglect of short-term profits	• Balance of short- and long-term profits
Growth of sales	• Aggressive capital investment	• Low cut-off rate for investment	• Higher cut-off rate
	• Global vision	• Overcapacity	• Restructuring • ROE, EVA, cash flow
Respect for people	• High morale and hard work create high-quality products	• Neglect of shareholder value	• Continued respect for people
	• Learning organisation	• Low growth, high labour costs, inflexible labour turnover	• More flexible personnel management, employability
Stability through group formation	• Alternative use of competition and alliance	• Low equity ratio	• Reduction of bad assets
	• Protection against takeover	• Low ROE	• Better financial health
As with the mission, emphasis on devotion to society	• Public trust in company and quality of products	• Environmental problems • Some unethical behaviour	• Social and environmental responsibility • Corporate ethics code
Philosophy and vision	• Sharing goals and meaning of jobs is understood	• Obsolescence	• Occasional updating
	• Consistent strategic decisions	• Not internalised	• Revision not top-down, but by a team

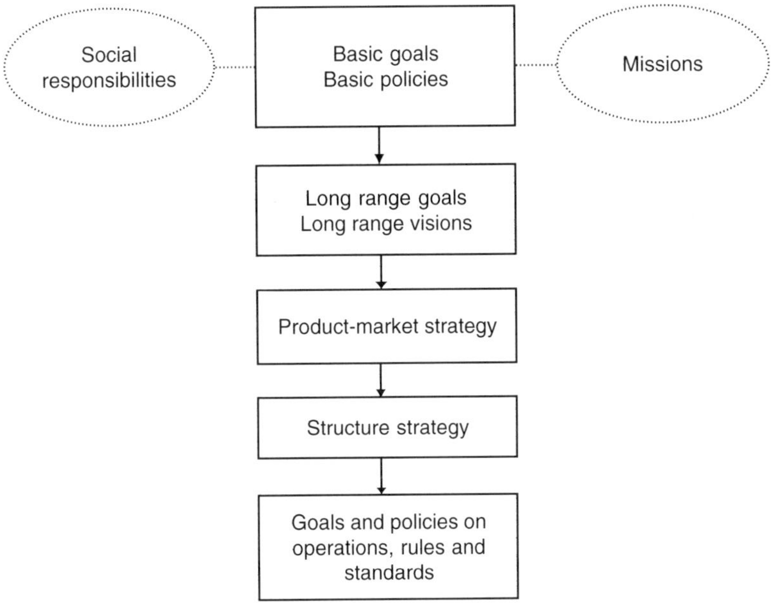

Note: Social responsibilities and missions are auxiliaries to the goal hierarchy

Figure 3.1 Categories of corporate goals

not to try to monopolise the world market are examples of social responsibility. Social responsibilities may reduce short-term profits but they should aid long-term sustainability.

The corporate philosophy seeks to capture the 'essence' or 'spirit' of the company's goals and strategies. Doing so often involves the use of a vivid image or a poetic metaphor, for example 'Sony thrives on exploring the unknown'. While a corporate philosophy may be implicit or explicit, a corporate creed is an explicit statement of the corporate philosophy. Figure 3.2 and Table 3.2 provide examples of the corporate creeds of successful companies.

A corporate creed should be short enough for every member of the organisation to remember it. It may be used at company meetings or on ceremonial occasions. Also, the words used should have an immediate emotional appeal to the organisation's members, as in the case of Sony, where everyone knows that being a member means 'thriving on exploring the unknown'. While the corporate vision (Table 3.2) appears to be similar to the corporate creed it tends to be

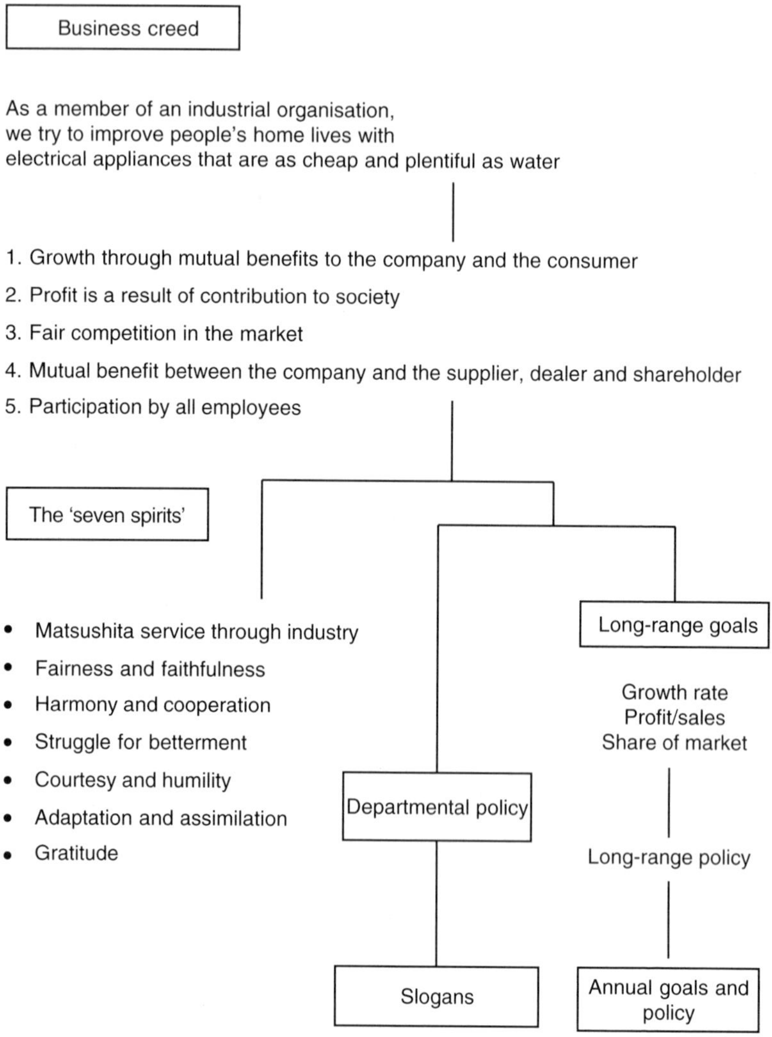

Figure 3.2 Business creed, basic goals and policies of Matsushita Electric

Table 3.2 The corporate creed and vision of successful companies

Corporate creed:

Hitachi's Odaira spirit
- Devotion to society, quality products for society
- Frontier spirit, self-help and aggressiveness
- Harmony and friendliness

The Sony spirit
- Sony thrives on exploring the unknown
- As an explorer, Sony opens up its own paths of progress
- Scientific and technological innovation are the vehicles of progress through which Sony hopes to serve mankind
- The tasks are not easy, but all members of the Sony family pull together to overcome difficulties, finding joy in creative work and pride in contributing their talents
- Sony believes in each member of its family, encourages the development of each person's abilities, and seeks the right place for the right person, to draw out everyone's best
- In this spirit, the united family is the powerhouse of Sony

Honda's management policy
- Always proceed with ambition and a youthful spirit
- Respect sound theory, develop fresh ideas, and make the most effective use of time

NEC's management philosophy
Through C & C [computers and communication] NEC
- tries to improve understanding between peoples of the world
- tries to help bring about an affluent world where people can fully express their personalities

Corporate vision:

Kirin Brewery
- Sales will be increased from 1250 billion yen in 1990 to 1700 billion yen in 2000
- Non-beer products will be 60 per cent of sales
- Diversity into biotechnology, construction engineering, information systems and the service sector
- The company will become globally excellent

more quantitative, defining goals, time schedules, product mix, future focus and image.

Perrow (1970) identifies several categories of goals and we shall follow his classification: (1) systemic goals, for example growth, stability and profit; (2) basic goals and policies, including structural and operational: policies; (3) output goals, defined by such phenomena as product-market strategy; (4) product goals, including quality, quantity and variety; (5) social goals, which address the corporate mission; and (6) derived social goals, such as achieving the corporate mission in a socially positive and environmentally sustainable way. Other dimensions include level of aspiration and time horizon (Richard, 1978).

Japanese corporations use their corporate goals and philosophy to encourage employees to identify themselves with the corporation. Hence they play an important role in securing organisational consent and organisational control. While detailed rules and standards are commonplace in American corporations, Japanese corporations seek to make the meaning and value of jobs clear through their corporate goals – by doing so, it is assumed, the hearts and minds of the employees will be engaged. Thus they try to create a culture that combines elements of moral involvement with an instrumental involvement but without alienation.

In conclusion we may note that, in general, goals and policies express the values of an organisation, help to guide decision-making and act as motivational aides that capture the meaning of jobs. They also provide a standard for evaluating the performance of individuals and departments (Richard, 1978).

JAPANESE CORPORATE GOALS

In this section we shall examine the corporate goals listed in Table 3.1.

Long-term vision

Japanese companies emphasise long-term goals over short-term profits. This is evident in the widespread practice of long-term planning and considerable investment in employee training, as well as extensive research and development to develop core competencies for the future. Hitachi and Sony spend about 10 per cent of their sales receipts on R&D, and this is not likely to change in the future.

Growth of sales

Growth and expansion were important values in the past because they increased the number of posts available for employees, and were necessary goals in the expanding economy that prevailed during the period of postwar development if the competitive position of a company was to be maintained. One can see this in the aggressive capital investment that occurred in the steel, car and semiconductor industries. The rate of growth increased with entry into foreign markets and foreign-based production as Japanese firms developed a global vision and foreign direct investment grew. The cut-off profit rate for investments remained low, the rate of return by DCF (discounted cash flow) rarely being computed. It was this that led to the overcapacity of the 1990s, which in part contributed to the Japanese economic recession. In 1996 the domestic production of cars was about 10 million, which was about 20 per cent more than consumption. Today the trend is towards setting a higher hurdle rate and putting more emphasis on profit as shareholders' interests are becoming more important with the increase in the power of these stakeholders.

Respect for people

Respect for people is evident in various personnel management practices, such as job rotation, the status ladder system and job security. The high morale created by this system contributes significantly to the quality of products. Because of the emphasis on training and the long-term accumulation of knowledge through lifetime employment and the effective transmission of knowledge across departments, many Japanese enterprises may be characterised as learning organisations. However the system also results in high labour costs because wage increases are partly based on length of service. Furthermore job security impedes labour flexibility, restricting movement to growing industries from declining branches of industry and businesses within them. The result is overemployment. Respect for people, while a desirable value, in the short-term tends to lead to a neglect of shareholder value and so it is often seen as a corollary of a low evaluation on the stock market.

The characterisation of Japan as a country in which there is lifetime employment in the core and peripheral labour markets is no longer accurate, and there is a growing tendency to emphasise capability and performance when determining wages and promotion. While there is increased emphasis on shareholder value, this has not yet meant the sacrifice of employee welfare. Rather it has become another part of the bundle of stakeholder interests represented in the strategic management of the enterprise.

Stability through group formation

Stability is an important organisational value, usually attained by appropriate equity ratios and a high rate of return, so that the company can continue to procure resources. In addition to this, many Japanese companies organise themselves into groups for increased security. The main bank system, the *zaibatsu* groupings, the *keiretsu* and cross-shareholding among group members are examples of this. Recently, low equity ratios and low returns on investment (ROE) have fallen out of favour in the stock market. We can see this in sudden declines in stock prices, which hamper the operations of companies, and in low credit standing, which makes it difficult to procure needed resources. The new mood in the Japanese market is to improve the financial health of enterprises – the fat of the growth era has been pared away, leaving companies more exposed and with less slack to absorb adverse shocks.

Devotion to society

Like individuals who sacrifice their private lives and devote themselves to the organisation throughout their working lives, Japanese organisations strive to show their devotion to Japanese society in general, not just their shareholders, expecting that this will result in enhanced public trust for the company. When national policy emphasised production rather than consumption it was in support of this belief: through growth, Japanese enterprises would restore to the nation the place in the sun that had been lost through the defeat and destruction of the Second World War.

Not all Japanese corporations have lived up to their mission statements – there have been disclosures of systematic malpractice and malfeasance by Japanese political and business organisations. After the bursting of the bubble economy of the late 1980s, unethical behaviour by a number of top managers became evident. The sudden shrinkage in the value of massively overinflated property portfolios exposed the absence of controls that had existed, especially in the financial sector.

It was also realised that Japan's environmental problems were a product of high economic growth. Consequently some of the corollaries of growth at all cost came under review. In the past, as long as these corollaries were externalities and were not charged to the corporate account – such as traffic congestion, intense air pollution and the noise made by the thousands of subcontractors who provide parts to the large corporate firms on a 'just-in-time' basis – little notice was taken of them. But when stakeholder models are broadened to acknowledge a range of environmental interests, then such externalities become harder to shrug off.

Philosophy and vision

Many companies have very specific corporate creeds, updating them when they become obsolescent. For instance Hitachi currently emphasises *quality* products, Sony emphasises *creativity* in innovation, while Honda emphasises *youthfulness* in the appeal of its products. The creed characterises both the corporate culture that the company wishes to sustain and the corporate strategy that it seeks to follow. The creed is a condensation of what it means to share common goals (in quality, in creativity, in youthful appeal and so on). The creed also makes clear the meaning of jobs in the enterprise: as oriented overwhelmingly towards quality, creativity or innovation, or towards the youth market. The creed, as the implicit backdrop to all strategic thinking, also encourages consistency in strategic decision-making.

CORPORATE PHILOSOPHIES AND CREEDS

Matsushita Electric

Clear communication of the company's mission and goals to its employees enhances their sense of identification with the organisation and guides their behaviour in the desired direction. The interesting case of Matsushita Electric serves as a demonstration of this. Matsushita has a clear statement of corporate philosophy, as shown in Figure 3.2. The creed uses a 'water supply' metaphor as its basic principle, one that involves mutual cooperation as well as emphasing profit. It also infers that the company will not engage in price wars. This business philosophy is developed further into the 'seven spirits', which all employees recite each morning at their work place – in this way Matsushita strives to indoctrinate these principles throughout the company.

The overarching business philosophy affects both the corporate culture and the strategy of the company. Because of its creed the company is marketing-oriented and is sometimes referred to as 'Matsushita the merchant'. Successful new products and the operation of foreign manufacturing plants owe a great deal to the sharing of these common values. The creed is constantly repeated in both word and print and employees are expected to exemplity it through their corporate deeds. The creed is elaborated in detail in the introductory booklet handed to new employees: the booklet explains the history of the company, its products, its organisational structure and the policy of each function. It is expected that all employees will want to perform in the best way to attain the goals of the company, and that, armed with this knowledge of the company, they will show initiative in performing their duties despite not being equipped with detailed job descriptions.

Each day begins with a morning meeting, where the company song is sung and the seven spirits recited before one employee gives his or her opinions about their work to the assembled organisational members. The text of the seven spirits is hung on the wall and the slogan for the current year is hung overhead throughout the plant, *kanban*-style.

These means of communication and indoctrination are practiced not only in Japan but also, with some modification, in subsidiaries abroad. Plant managers typically respond to the question 'What do you consider to be the most important work motive for ordinary female employees?' by suggesting that the user's satisfaction with Matsushita products is one of the most important drivers of employee motivation. The message that the customer is sovereign has been well and truly inculcated throughout Matsushita's plants.

The practices of most successful Japanese corporations are close to those of Matsushita. Toyota, Hitachi, Canon and Honda all have formal business creeds and their business goals are communicated in a number of typical ways. These include a 30–150 page induction and guidance book for new employees, a monthly or fortnightly company magazine, slogans, morning meetings, training programmes, a New Year's message from the president, and abstracts of long-term plans and budgets (see the example from Kirin Brewery in Table 3.2).

Survey of corporate creeds

Table 3.3 shows the results of three surveys on the business creeds of about 300 companies, conducted in 1964, 1986 and 1998. The creeds consist of basic goals and values, strategic and general policies, and codes of behaviour towards the company, work, seniors (or authority figures), colleagues, subordinates and oneself. The six most frequently cited items were good products and a customer orientation, contribution to society, innovation, creativity, faithfulness, and cooperation.

The changes in the frequency of the items between the periods reflect changing company values and strategies. A comparison shows that a number of items are not as valued nowadays as they were in the 1960s, including the notion that good products develop customer focus and contribute to the nation and society. Increased importance has been given to progressiveness, innovation and creativity. The development of desirable attitudes has diminished in importance, but importance has continued to be attached to cooperation and the building of mutual trust. Overall, companies place great emphasis on values such as customer orientation, innovation, creativeness and mutual cooperation.

Success factors in corporate philosophy

Many authors in the West have stressed the importance of corporate philosophy. Best-selling books such as *In Search of Excellence* Peters and Waterman, 1982), *Built to Last* (Collins and Porras, 1994) and influential articles such as 'Strategic Intent' (Hamel and Prahalad, 1989) are just a few of the many texts that are influenced by Japanese practice but whose focus, with the exception of Peters and Waterman, is all-American. Peters and Waterman were clearly responding to the emergence of Japanese practices as iconic for US business in the early 1980s and were seeking to develop some home-grown exemplars for local (and indeed global), consumption.

Here we shall examine the conditions of success. How can overall policy guide correct decision-making? Why has Sony's creed – 'Sony thrives on exploring the unknown' – led to the creation of successful new products, such as the new type of colour TV (Trinitron) and the Walkman (total production amounted 170 million units in 1999)? How has Honda's creed – 'Always proceed with ambition and a youthful spirit' – ensured the successful production of motorcycles and cars?

Table 3.3 Survey of corporate creeds

	1964	*1986*	*1998*	*Change (over 20 cases)*[1]	*1998 bar frequency*
Number of companies[2]	269	300	270		270
Basic goals:					
Good products at reasonable price	112	69	70		———
Customer orientation, needs orientation	19	58	104	+	————
Customer value creation	–	–	41	+	——
Contribution to the nation	87	19	8	–	–
Contribution to society	44	145	171	+	—————
Mutual prosperity	18	17	23		—
Prosperity of the company	13	55	67		———
International vision	–	37	51		——
Fair profits	9	11	14		–
Steady progress	22	10	16		–
Gaining trust	17	28	53	+	——
Welfare of employees	12	40	40		——
Respect for employees	15	32	61	+	———
Responsibility to stockholders	5	7	13		–
Social and environmental responsibility	–	–	69	+	———
Strategic and general policies:					
Progressiveness, action-orientation, innovations, modernisation	50	71	94	+	————
Analytical scientific attitude	44	33	2	–	–
Strengthening marketing capability	20	5	4		–
Creativity	42	85	38		——
Unique products	7	13	19		–
Technological innovation Improvement of research and development capability and high level of technology	78	42	53		———
Refined internal organisation	5	4	12		–
Human resource cultivation	11	11	20		—
High productivity	32	39	10	–	–

Table 3.3 *Continued*

Codes of behaviour:					
Attitude towards company					
Faithfulness	81	68	39	–	——
Gratitude	22	14	16		–
Pride in company, respect for tradition	13	25	22		—
Attitude towards work					
Effort, patience	53	34	23		—
Vitality, courage, fighting spirit	–	36	41		——
Responsibility, fulfilment of duty	45	39	11	–	–
Observation of rules, correct judgement, business ethics	16	5	6		–
Thrift, cleanliness	29	6	3		
Cheerfulness	–	9	6		
Attitude toward seniors, colleagues and subordinates					
Cooperation, cohesiveness	145	113	43	–	——
Mutual trust	7	38	11	–	–
Kindness, generosity	25	5	1		–
Love	18	12	2		–
Fairness	7	13	10		–
Courtesy	25	17	9		–
Attitude towards self					
Health	28	18	3		–
Diligence	15	15	7		–
Cheerfulness	26	19	4		–
Self-training, self-realisation, dreams, hopes	47	46	36		——

Notes:
1. Long-term trend.
2. Frequency (number of companies that have each creed item).
Sources: Oyama (1965); Japan Productivity Center (1986, 1998).

Part of the answer lies in the emphasis these firms give to the 5S principle in respect of quality improvement: *Seiri* (putting in order), *Seiton* (arranging properly), *Seiketsu* (cleanliness), *Seiso* (cleaning) and *Shitsuke* (good behaviour). The 5S slogan is seen everywhere in the plants of Japanese corporations.

We believe that two assumptions underlie the widespread adoption of the philosophy and slogan. First, if based on appropriate experience and knowledge, an overall policy can lead to good decisions on the details of its application. For instance when one makes a drawing or painting one starts with a rough sketch or the 'design', with the details in one corner, because the whole is more important than the details. When one constructs a house, one starts from the concept upon which the design of the new house will be premised. Successful leaders of new product development say that the 'concept' of the new product is the key to success, and in order to find a potentially successful concept firms conduct market research and explore new ideas creatively and imaginatively. But there are also appropriate principles for decisions. There are several principles in Japanese art that lead to a good and enjoyable picture, such as composition and harmonious and beautiful colours. In management the principles that underlie management practices fulfil a similar function. Creative and successful management principles can be crystallised into the corporate philosophy or vision. The corporate creeds of Hitachi, Matsushita, Sony and Honda are all based on the successful experiences of their founders. Yet recent trends suggest that one should instead form a team to study the corporate creed, and after collecting sufficient performance data, the corporate philosophy and vision should be stipulated.

There are three sets of principles that seem to be common to successful corporate creeds, all of which promote the interest' of stakeholders: (1) customer orientation and superior customer service; (2) progressiveness, innovation and the pursuit of excellence; and (3) respect for employees and individuals. Hitachi, Sony, Honda and Canon all emphasise these principles, which can also be found in much of the management literature (for example Peters and Waterman, 1982; Kotter and Heskett, 1992; Wilson, 1992).

Failed creeds

To understand the effectiveness of creeds we have studied some cases of failure. The Ataka Trading Company was founded in 1904 and was the tenth largest trading company in Japan (sales in 1975 were 2000 billion yen). However it went bankrupt when a newly established oil refining company in Canada broke a large trade agreement on oil sales. That company went bankrupt in 1975, owing Ataka 100 billion yen. The top management of Ataka remained unaware of the loss for almost a year, because the director in charge of the contract hid the facts. The company was controlled by the Ataka family; it had neither an appropriate top management organisation nor a business philosophy. The owner seemed more interested in collecting fine art and classic cars than in running the company (he had 42 classic cars!) This appears to be a clear case of a lack of corporate philosophy and corporate control.

The Eidai Company, which produced laminated board and constructed housing, had sales of over 100 billion yen in 1977. However it went bankrupt in 1980 because of overexpansion. The corporate creed was as follows:

- Apply creative ideas by using your brain.
- Those who do not have creative ideas should work their bodies hard, into a sweat.
- Those who cannot use either their brain or their body should leave the company quietly.

The problem with this creed, which was displayed throughout the company, was its evident lack of respect for people, implying that many were stupid and incapable of using their brains. Nor was the creed customer-oriented. The president, an autocrat, adopted an extreme top-down decision-making style.

Van Jacket was a fast-growing company producing suits and sweaters for young people, including an 'Ivy League Look' suit with metal buttons. The suits sold widely and the president, a fashion designer and artist, wrote many articles for newspapers and journals. Sales reached a peak of more than 45 billion yen in 1975, but the company failed to foresee a change in fashion among young people, based on more informal patterns and designs, and it accumulated a huge stockpile of unsold suits. Its designs ceased to be popular when it lost its ability to interpret street-fashion. Moreover the company did not have a corporate creed – the business philosophy of the president was 'enjoy yourself'. The culture of the company was oriented solely towards fashion but it lost the plot in terms of youthful styles, ignored risks, lacked a customer orientation and took no heed of market signals. Above all the company lacked interest in financial management, its equity ratio being only 1 per cent. This case demonstrates company failure due to the lack of a corporate creed and the failure of corporate philosophy to follow good principles.

Revision of creeds and visions

The recent trend is for the creed and vision to be updated every three to five years as changes occur in the environment and in the values and power of stakeholders and interest groups – otherwise the business creed becomes obsolete. Another recent trend in the formulation of new creeds and visions is to use project teams of young managers. Young managers have longer-term expectations for the company because they will be there longer than the higher level managers. Their longer-term future in the company thus makes them more committed, materially, to the future of the company. They also collect and analyse information that may come to replace the experiences of founders. It is absolutely necessary to do this: the wisdom of age does not necessarily endure in a fast-moving business environment.

CONTEMPORARY TRENDS

Social and environmental responsibility

Social responsibility involves responding to the needs of the larger society and natural environment while sacrificing short-term profits (Kono, 1999). It is different from a product-market strategy in the sense that it does not aim just at making a profit, in fact it reduces short-term profits. Social responsibility extends the activities of the corporation into areas that affect not only society but also the natural environment. The market mechanism does not necessarily work here: it is always possible for a company to ignore serious environmental risks in order to make a short-term profit. But the risks are immense if not managed properly. Such companies receive unexpected social feedback in the form of law suits, unwanted government intervention, regulation and control, public action campaigns, criticism from the media and a massive depreciation of whatever social capital is attached to the image of the business. For instance American tobacco manufacturers have had to pay $360 billion to some American states and some patients, and will continue paying for 25 years as a result of a successful 1977 lawsuit against them. In Japan the demand for polyvinyl chloride, which is used to make containers for various goods such as cosmetics, is falling because it produces a harmful gas when burned. This is a case of non-economic social feedback affecting the market mechanism.

No company can survive in the long term without paying attention to social responsibility. In Japan four newspapers – *Asahi*, *Mainichi*, *Yomiuri* and *Nihon Keizai* – are particularly powerful and have a considerable daily circulation (*Asahi*, 8.3 million, *Mainichi*, 3.9 million, *Yomiuri*, 10.1 million and *Nihon Keizai*, 3.0 million in 1998). If they report on environmental pollution or other unethical behaviour by companies the sales and profits of these companies are eventually affected. Monitoring by major newspapers and other media works as both an early warning system on, and a fact-finding system for, the activities of corporations.

There are three areas of social responsibility, according to the level of duty. Corporate philanthropy comprises activities that are deemed desirable for a corporation to engage in. Corporations deal with a large amount of resources and if they have surplus resources they can help non-governmental organisations and other organisations that operate tangentially to the market mechanism in their mission, such as educational activities, research or public services.

In a narrow sense social and environmental responsibilities are related to 'should-do' or 'should-not-do' activities. As the activities of corporations expand they affect the social and natural environment. Thus organisations have to analyse the effects of their corporate activities and respond to social and natural environmental needs through risk management and avoidance. Such management and avoidance can occur only through the sacrifice of short-term profits.

Corporate ethics deal with 'must-not-do' or 'never-do' issues. They are prohibitions against violating laws and social duties. Japanese corporations try to advance social welfare, as can be seen in the corporate creed in Figure 3.2. On the other hand, because of the production-first ethos and company members' devotion to the corporation, some unethical behaviour does occur, such as concealing corporate financial losses or giving large amounts of money to gangsters who threaten to disclose dishonest financial statements at the shareholders' meeting. It is to try to minimise the risk of gangster involvement that many companies tend to hold their shareholders' meetings at the same time on the same day.

Social and environmental responsibility has come to be important for a number of reasons. The volume of activities by manufacturing companies has expanded greatly, and with this expansion the potential impact of their activities on society and the natural environment have also increased.

Table 3.4 A broad classification of social responsibility and required level of behaviour

Classification	*Level of required behaviour*	*Examples*	*Parties affected*	*Direct relationship with main business*
Philanthropy	Desirable to do	Help for NGOs, voluntary donations	Mostly social environment	No
Social responsibility (narrower sense), environmental responsibility	Should do	Employment of disabled people, recycling of products, pollution prevention	Consumers, suppliers and dealers, employees, society, natural environment	Yes
Corporate ethics	Should never do	Collusion, bribery, manipulation of financial statement	All environments	Yes

Table 3.5 Examples of a standard of social responsibility index

To whom responsible	*Item*	*Level of standard*
Consumers	Cost of production/retail price	60% or more than the industry average
	Maximum share of market	30–60%
Wider public domain of corporate environmental impact	Use of freon gas	None
	Percentage of recycling	More than 50%
	Residues from operations	Reduction by 50%
	Power consumption	Reduction by 50%
	Energy for distribution	Reduction by 40%
	Index for the greening of industry, at a plant level.	Percentage index of green zones, with minimum amounts of emissions, noise, accidents
	Donation	2% of income or sales
Employees	Assurance of employment	Assure employment of regular employees, except in employment crises
	Employment of disabled people	3% of all employees
	Employment of older people	Average percentage of 60–70 year old people in the population
	Accident rate	Average of frequency and severity in other companies, or zero
	Expenditure on employee welfare over and above legal necessity	8% of labour cost (plus legally required expenditure)
Shareholders	Disclosure of information	Set rules on disclosure of information

The main areas of social and environmental responsibility are shown in Table 3.4, and Table 3.5 presents an example of a standard of social responsibility index. The subjects to whom the responsibilities are addressed are consumers, those in the wider public domain of corporate environmental impact, employees and shareholders. The activities of corporations affect these sectors both positively and negatively, but feedback is often defective. Positives do not necessarily entail profits to the company and negatives do not necessarily entail losses to the company, because both the market mechanism and the social feedback systems are imperfect. However, eventually there may be extensive and costly feedback. For example the company Nihon Chisso had to pay about 500 billion yen to people who had eaten fish poisoned by the mercury that the company had unwittingly discharged into the sea. Showa Denko, a large pharmaceuticals manifacturer, sold a drug in the US (Triptofan, an amino acid drug) that caused serious poisoning, and the company had to pay about 100 billion yen between 1990 and 1995 to patients who won a product liability suit. Neither company had knowingly caused harm, but innocence of intent is no excuse for escaping indemnity. Companies have to make every effort to assess the effects of their activities.

Life-cycle assessment

Companies have to assess the effects of their activities throughout their products' lifetime: from the procurement or production of raw materials, through assembly, distribution, sale, use by consumers and disposal. Until recently companies were not required to consider the effects of their products on the environment after a sale was completed, but now they are responsible for the recycling of used packaging and products. The ISO14000 system requires companies to assess the effects they have on the environment, and possession of an Eco Mark is fast becoming a requirement for companies wishing to export their products abroad. A very high percentage of Japanese corporations have obtained a license from the ISO for all their overseas plants. In 1999 Matsushita won an ISO14001 license for 101 plants in Japan and 120 plants elsewhere in the world (ISO14000 emphasises environmental management and does not refer particularly to life-cycle assessment; on the ISO14000 see Uetani, 1997).

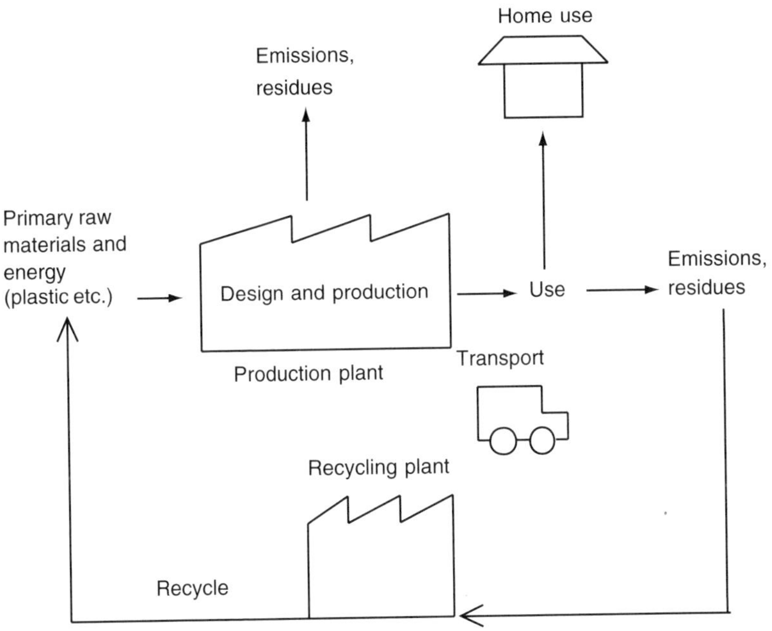

Note: The impact on the environment of each stage from designing to recycling is measured by the amount of waste generated.

Figure 3.3 Life-cycle assessment

Decision-making principles

Three main management principles pertain when deciding how to respond to social and environmental needs: the maximising principle, the dominating principle and the satisficing principle (on satisficing see March and Simon, 1958; on the dominating principle see Ruce and Raiffa, 1957).

The *maximising principle* assumes that private companies can maximise social welfare. But they compete on quality and cost of products and cannot survive by trying to maximise social welfare. It is difficult for organisations to be maximisers with respect to welfare: how can they construct a unitary utility schedule by using the conversion rates of various effects for social welfare?

When applying the *dominating principle* companies seek to reduce costs and increase sales while at the same time meeting their social and environmental obligations. For example reducing the consumption of energy and materials reduces costs; locating the factory in a rural area reduces labour costs, as does employing disabled people. Parts and packagings can be recycled to reduce material costs and reusable materials are favoured (for instance glass bottles are easier to recycle than plastic bottles). All these actions can serve wider social and ecological purposes.

Life-cycle assessment can be useful in identifying options in other aspects of planning. Sometimes this may be an aspect that was not evident when the product was initially designed. For instance Fuji Film sold the first single-use camera (*Utsurun-desu*, or Quick-snap) in 1986 and it was an instant success. However when it was launched in the US in 1987 it failed to take off because consumers doubted its reliability and environmentalists denounced it as a generator of waste. The camera was redesigned to be recyclable and by 1992 it had attained a 20 per cent share of the film market. All such cameras are taken to film processors for development so recycling is easily arranged – about 90 per cent are recycled and the decomposition and selection of used parts is automated.

Fuji-Xerox Company in Japan invented a recycling system called the 'closed loop system'. The company recycles not only toner cartridges but also every part of the copiers it produces and services. From the initial design of copiers, parts are planned with an eye to making decomposition much easier. In the car industry it is no longer just a question of cost reduction as a condition for survival, firms also have to be able to offer a wide array of environmental features, such as reduced petrol consumption, reduced emissions and safer bodywork.

The *satisficing principle* means that a company does not maximise social welfare but sets minimum social welfare standards while at the same time maximising profit and sales. For example airborne emissions and water borne waste from a factory are not cleaned up by the firm to the level where the air can be breathed or the water drunk. They can be cleaned after leaving the organisation by the natural ecological system. A company need not employ the maximum number of disabled workers but it should ensure that up to 3 per cent of all employees are registered disabled, the average ratio in Japan. By establishing satisficing standards, a company can strive to make efficient use of resources in its main business.

In Table 3.5 we presented a standard of social responsibility index, which was consolidated from a number of case studies of Japanese corporations. There are three approaches to establishing a standard.

The first of these is the *effects–cost ratio approach*. For instance the cleaner the emissions from a car engines the greater the social welfare, but there is also an increase in the cost of emission control. At some point the cost increases will become disproportionate and grow more rapidly. It is this point that is used for a standard. Once past this level the wastage of natural resources arises. This can be improved by scientific or design innovation. While this approach is the most rational, the measurement of the standard point may be difficult: sometimes other proxy variables may be required.

A second approach involves the natural ecological system. Many ecological systems have the power to recover their balance. An increase in atmospheric CO_2 may be accompanied by an increase in trees to absorb it. CO_2 is also absorbed by the sea and corals. To some extent rain can clean polluted air. Thus factories do not have to take maximum measures because of the cleaning power of nature. Nature also has restorative power. Trees and fishes can be restored, as can the water that feeds hydro-electric dams. Such restoration may provide the basis for a standard.

Another way of setting a standard is through the 'average approach'. The average percentage of a phenomenon in the broader society may be fixed as the company standard for that phenomenon. For instance the employment ratio of disabled to able-bodied workers can be derived from the average proportion of disabled people in the broader society. Similarly the ratio of donations from profit-making organisations to non-profit-making causes can be fixed by considering the average donation by companies.

A final approach to standard setting is to fix a gradual improvement rate, whereby companies move progressively towards a fixed target, for example to achieve a 40 per cent reduction in energy consumption or 50 per cent recycling of components. The advantage of this approach is that, rather than just making an arbitrary effort to improve a standard, the goal is clear to the company.

Corporate ethics

Recently there have been many instances of unethical behaviour by management, resulting in bankruptcy or forced retirement for the top managers. Thus corporate ethics have become a focus of attention. The bankruptcy of Yamaichi Securities and Takushoku Bank are typical cases. Their losses were caused by the fall in the price of land and securities after 1990. Financial statements were manipulated to hide the losses, and in an attempt to make up the short-falls unsuccessful speculative investments were carried out, thus compounding the losses. In other cases bribes were given to gangsters (*Sokaiya*) to prevent them from disclosing the losses at the shareholders' meeting.

Corporate ethics means complying with the norms or laws of society and consciously seeking to ensure that the business and its representatives do not engage in illegal or unfair practices. It is a collection of 'must-not' rather than 'must' prescriptions (Table 3.6). There are three arenas at which unethical behaviour is targeted: the primary environment or the market, the secondary environment, and the stakeholders.

Three categories of unethical behaviour can be identified (Robin *et al.*, 1989): behaviour that is illegal, unfair or violates social norms (most examples of unethical corporate behaviour belong to this category); social welfare utility, where corporate ethics are judged in terms of their disutility being greater than their utility, (for example insider trading of stocks or purchasing from friends); and breaches of trust (such as the sale of defective goods or misleading advertising).

Unethical behaviour may arise from an overly developed sense of loyalty to the company. In the case of Yamaichi Securities and other bankrupt companies the top management thought that if the large losses by the companies were disclosed the companies would not be able to survive and employees would lose their jobs, so they doctored their financial statements. Keeping their posts in top management was a secondary reason. Likewise, the antimonopoly law may be violated in order to increase sales and profits. Unethical behaviour at the top management level also comes from a notion of duty of care that overwhelms any sense of ethical duty. In Japan, boards of directors are composed of executives of the company in question, which makes it easy for top managers to conceal unethical behaviour.

Insider accusations and whistle-blowing are rare in Japanese corporations, where friendship is important to long-term employment. A whistle-blower may be afraid of his accusation being turned against him because 'flexible' behaviour is sometimes regarded as normal. For example minor violations of the antimonopoly law are regarded as acceptable. Informal discussions with a competitor on a bidding price is another case in point (see Bird and Walters, 1989).

A recent development has been the establishment of codes of ethics. Few Japanese companies had such codes in the past because it was assumed that employees would be honest and loyal (examples of codes contained in corporate creeds can be found in the section on 'codes of

Table 3.6 Unethical business practices

Target group	*Unethical practices*
Consumers	Sale of harmful or defective products, false or misleading advertising, cheating
Suppliers and dealers	Unfair trade, list price control
Competitors	Cartel formation, collusion, division of markets, unlawful differentiation, dumping, use of patents to limit competition, bribery, illegal discounts
Natural environment	Pollution, environmental destruction
Government	Tax evasion, bribery, illegal political contributions', evasion of reporting, false reporting, revealing breaches of laws
Community	Industrial accident, industrial pollution, illegal dumping, planned bankruptcy
Employees	Lack of attention to workplace safety and occupational injuries or disease, causing death by overworking, employment discrimination, sexual harassment
Investors	Insider trading, bribing gangsters, indemnifying selected customers from loss at the sacrifice of other customers, manipulation of share price, window-dressing of financial statement
International	Use of tax havens, price dumping, political manipulation

Source: Adapted from Nakamura (1997).

behaviour' in Table 3.3). Corporate creeds, typically state 'do this and that' rather than 'don't do this or that'. Shiseido (a cosmetics manufacturer) enacted the Shiseido Code in 1997, containing almost all items in Table 3.3.

Changes in the mode of selection of non-executive directors are hastening ethical disclosure and compliance. Sony has reduced the size of its board from 38 members to 10, three of whom are non-executive directors. This measure is spreading to many other companies; thus the disclosure of company activities will be advanced. Furthermore the new code in the Company Act stipulates that at least one member of the statutory auditors must come from outside the company, and this auditor could act as a monitor of corporate ethics. At Shiseido an independent centre for corporate ethics has been established.

Shareholder value

Shareholder value has not necessarily been neglected in the past. When the economy boomed, profit growth accompanied the growth of sales and share prices rose. In 1983 the price of major stock was 9893 yen (Nikkei average) but by 1989 it had risen to 38 919 yen. Even though it fell to about 17 000 yen between 1992 and 1999, this was still twice the 1983 level. However the stock prices of most companies have fallen since 1995 and those of many financial institutions and construction companies have fallen to one tenth of their previous value, a precipitous decline that has threatened the survival of many of these companies. On the other hand the average proportion of holdings by foreign investors has increased to 10 per cent for all stocks listed on the Tokyo stock exchange, and for some companies the figure is more than 30 per cent.

To date, banks have tended to adopt a long-term view with regard to corporate borrowers and have not interfered with their management. Likewise, neither shareholders' meetings nor boards of directors have functioned as controlling organs for shareholders. But since the mid 1990s the stock market has had a more powerful influence, encouraging management to pay attention to profits.

Return on equity

Return on equity is measured by profit after interest and tax divided by equity or net worth:

$$X = r(E + D) - kD \ldots \text{net profit}$$

$$(X/S)/k = r + (r - k)D/E \ldots \text{ROE}$$

$$P = (X/S)/k \ldots \text{stock price}$$

Where X is the profit after interest and tax, r is the after-tax return on total assets before interest, P is the share price, X/S is earnings per share, D is debt, E is equity and k is the interest rate.

ROE is high, when (1) r is high, (2) $r > k$ and the debt–equity ratio (D/E) is high, or (3) when $r < k$ and D/E is low. This is the leverage effect. In Japan, r has been low and the leverage ratio (D/E) has been high. The ROE of Japanese manufacturing corporations has been very low (Table 3.7).

The average ROE of eight typical American corporations (Dow Chemical, Ely Lilly, IBM, Boeing, Philip Morris, Pepsico, Intel and GE) between 1994 and 1997 was 19.2 per cent. The comparatively low ROE of Japanese corporations results in a low share price because of the low earnings per share. This can be explained by means of a simple model.

$$X/S = \frac{X/S}{E/S} \ldots \text{ROE}$$

$$(= \text{earning per share divided by equity per share}) \quad (4.1)$$

$$P = (X/S)/k \ldots \text{share price} \quad (4.2)$$

where X/S is earning per share, E/S is equity per share, x is profit after interest and tax, E is equity, S is the number of stocks issued, and P is the share price. If E/S is constant, then high X/E results in high X/S (Equation 4.1), and high X/S results in high share price (P) (Equation 4.2).

Table 3.7 ROE of 373 Japanese manufacturing companies, 1991–95 (per cent)

1991	*1992*	*1993*	*1994*	*1995*
5.44	2.94	2.13	2.53	4.64

Source: Nikko Research, 1997.

EVA (economic value added) and MVA (market value added)

In addition to ROE, many Japanese companies compute EVA and use it as one of the goals of the company and its divisions. EVA is net profit after deducting tax, interest and target return on equity:

$$\text{EVA} = Y - kD - rE$$

Where Y is profit after tax but before interest and dividends, k is the interest rate and r is the target rate for equity or cost of equity.

This is computed by determining risk minus interest rate plus risk factor (which we do not discuss in this book). We present two examples in Table 3.8. We can see how the EVA differs between the two companies, although Toray is one of the best companies in Japan. EVA is surplus profit and has the following properties. First, the share price rises if EVA is positive and falls if it is negative. This plus or minus measurement is a clear yardstick from the viewpoint of stockholders, thus pension funds in the US use it to evaluate companies. Second, it is a medium-term measurement. Low inventory reduces asset value and increases EVA. It lets the company raise the cut-off rate for capital investment, resulting in improved profitability by staying away from low-return investments. Third, it can be used to evaluate divisional performance. If EVA is positive the division can be expanded, if it is negative the division should be cut back. In this case the balance sheet has to be divided into divisional balance sheets.

MVA represents the total market price of stocks and bonds minus the original paid-in amount for stocks and bonds. It is affected by EVA. If the market value of the company is high, the company can float stocks and bonds at high prices and reduce the cost of capital (for EVA and MVA see Stewart, 1991; Grant, 1997).

There are some limitations to the use of ROE and EVA. First, they represent goals, and if the product-market strategy is not right these goals will not be attained. Second, they ignore long-term growth and can disturb long-term performance. R&D expenditure or necessary investment could be curtailed in order to contract capital assets and increase short-term profits. Third, they may not serve the interests of other stakeholders and interest groups. They may be used to inhibit wage increases, weaken employment rights, impede the improvement of process quality and the need to invest in environmental protection.

Table 3.8 Two examples of EVA

1. Toray (1996, unconsolidated, billion yen)
   Sales 543, equity 432, profit after interest and tax 17.8

   432 × (0.02 + 0.03) = 21.6 ... target cost for equity, 0.02 ... risk-less interest rate, 0.03 risk-factor (or standard deviation of ROE of the industry).

   EVA = 17.8 − 21.6 = −3.8 (million yen)
   (as reference, dividend ... 4.9, ROE = 17.8/432 = 0.04)

2. Intel (1997, consolidated, million dollars)
   Sales 25 070, net worth 19 295, net income after interest and tax 6945

   19 295 × (0.05 + 0.03) = 1543 ... cost of equity

   EVA = 6945 − 1543 = 5402 (million dollars)
   (as reference, dividend ... 180, ROE = 6945/19 294 = 0.36)

Cash flow is the net cash available for operational activities, investment and financing activities. If profits are higher than liquid assets, the net cash flow from operational activities will be large and the need for external financing will fall. Thus, the lower the growth rate, the higher the profit rate and the higher the net cash flow. The acquisition of other companies by exchange of stocks will not change the cash position, but LBO (leveraged buy-out) or MBO (management buy-out) will increase the debt and reduce the ability to borrow in the future.

The reasons why cash flow has grown in importance for Japanese companies since 1995 are as follows. First, bank lending had to be restricted because of the bad debts caused by the fall in land prices, hence many manufacturing companies have had to strengthen their self-financing capacity. Second, a high cash flow improves both the bond rating and the ability to obtain external finance. Third, it encourages financial independence among divisions, making them more careful about investment. The bad debts accrued by many construction companies and financial institutions were caused by ignoring cash flow. Fourth, since March 2000 Japanese companies have had to include cash flow in the consolidated financial statement. Finally, and most importantly, it is a measurement of financial security and companies now plan their investment decisions better by computing NPV (net present value), through computing the future cash flow, using the discounted cash flow method.

These new trends may be seen as evidence that Japanese corporations are becoming more American in style and oriented towards short-term profits. However we do not support this convergence interpretation. Instead we argue that the trend is towards balancing multiple goals, balancing the rewards to multiple stakeholders depending on changes in their relative power or the changing contribution of resources. Balancing the value of goals is a general principle of goal formation (Clarke and Clegg, 1998). The new trend will curb the tendency of some Japanese corporations to apply too low a hurdle rate for investments.

SUMMARY

Goals and policies are hierarchically arranged. Basic goals and policies are central values of corporations and are determined by the structure of corporate governance.

In Japan the power of shareholders has been weak but the influence of employees and banks has been strong; thus companies have tended to emphasise the long-term growth of sales. Consequently the cut-off profit rate for capital investment has been low, which was a successful policy during the period of high economic growth, but when the economy slowed it led to overcapacity and severe competition. The balance of goals is now changing.

Corporate philosophies and creeds contain goals, missions and social responsibilities and are worded to appeal to employees. The creed is sometimes recited at the morning meeting and is meant as device through which the members can share the values of the organisation and identify with the organisation. All the values contained in the creed, including respect for people, are expected to be accepted by employees and internalised by them. This is similar to Etzioni's category of moral involvement, but it is being built into commercial and market-based organisations rather than into values-based religious or similar organisations. Loyalty to society and the organisation is emphasised, as are customer orientation, progress and respect for people.

Among the contemporary trends in Japan are the establishment of a corporate code of ethics and an effort to promote the disclosure of information by selecting non-executive directors and outside statutory auditors.

There are two approaches to carrying out social and environmental responsibilities: the dominating principle and the satisficing principle, using attainable standards to balance multiple goals.

Another new trend is to place increased importance on shareholder value. To this end, ROE, EVA and cash flow have become more important. This does not mean that Japanese corporations are becoming more American style, but that the weight attached to various goals changes according to the power of stakeholders and changes in governance structure.

4 Product Mix and New Product Development

AREAS OF CORPORATE STRATEGY

A firm's product-market strategy focuses on the choice of product areas in which to operate, thus positioning the company in specific relationships to its operating environments. Product-market strategy has a considerable effect on company performance. When the demand for a product is declining, growth is impossible, no matter how sophisticated the organisational structure. There are four main areas in which product-market strategy can be chosen, as shown in Figure 4.1. The product mix reflects the market segments and value-premise that the company will use to attract its customers. Vertical integration fixes the boundary between the company system and the market: the extent of control by the company over the value chain. The degree of multinational management reflects the extent to which the production activity operates in other countries through the transfer of firm-specific management and technical skills. It is often assumed that these are superior to those available in the new country of operation. Competition strategy comprises the actions taken to hold or increase the company's share of the market.

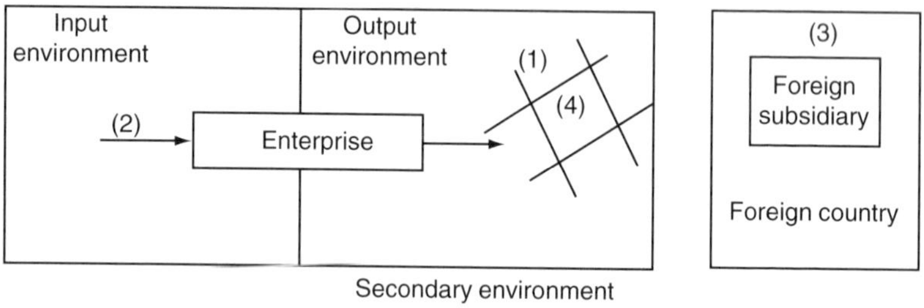

Notes:
1. Product-mix strategy
2. Vertical integration
3. Multinational management
4. Competition strategy

Figure 4.1 Four product-market strategies

These four strategies are complementary to some extent, but beyond a certain level they are independent and may compete for resources.

The objectives of this and the three subsequent chapters are threefold. First, we will classify different strategies and clarify the concepts that underpin them. Second, we will investigate the empirical effects of different strategies on performance in order to determine the principles governing the strategies and performance of Japanese companies. Third, we will identify the key characteristics of the strategies of Japanese enterprises, especially in light of recent trends.

The literature on corporate strategy is a growth area underlain by some pioneering works, which we shall review here. Early research was carried out by Gort (1962) on the diversification and integration of US companies. His approach was derived from the economics discipline and there was no analysis of the interconnections between strategy and structure. Perhaps the most famous of the pioneers were Ansoff (1965), whose *Corporate Strategy* was a pioneering work on strategic decision-making processes and the principles behind diversification, and Chandler (1962), whose classic work on strategy and structure inspired many other significant works. For instance Channon's (1973) research on the strategy and structure of British enterprises drew inspiration from Chandler, as did Rumelt's (1974) research on the strategy and structure of American enterprises; in part, each of these was also inspired by Gort. While Rumelt's research was less well developed in its analysis of vertical integration and multinational management and did not consider the process of strategic decision-making, Channon provided no analysis of the performance of each type of strategy. Rumelt's work influenced Yoshihara's Japanese management research (Yoshihara, 1986). While enlightening, this work did not analyse integration or multinational management. Kono's (1984) book on the strategy and structure of Japanese enterprises provided a more comprehensive analysis of strategy. Finally, as almost everywhere, Porter's (1985) book on strategy enriched Japanese research on competition.

THE CONCEPT OF PRODUCT MIX AND DIVERSIFICATION

A firm's product mix should consist of growth products that separate the performance of the company from the products' life cycle (Gort, 1962). The demand for any one product will grow and then decline. By systematically changing the product mix the company separates its own fate from the fate of its products. If the demand for one product continues to grow for many years, the company can specialise in that one product and continue to develop.

There are several ways of thinking about product mix. In Japan, classifications deriving from Channon (1973) and Rumelt (1974) have been influential and we use them here – but not in exactly the same way. Rumelt included vertical integration as an element of diversification but vertical integration differs from diversification. Diversification relates to the variety of uses of products and separates the performance of the company as a whole from the product life cycle. Rumelt also neglected marketing relationships, emphasising only technological relationships. His concepts of 'constrained', meaning limited to one core technology, and 'linked', meaning two or more core technologies, are not only complicated in their original application but also inappropriate for contemporary industrial synergies. Instead we consider the mutual relationship between products as a group.

A single-product company (S in Figure 4.2) is a company in which one product accounts for more than 95 per cent of total sales. For example 95 per cent of Toyota's total sales are of cars, so it should be classified as a single-product company. A dominant-product company (D) has one product that accounts for 70–95 per cent of total sales. For example, for Asahi Brewery beer accounts for about 80 per cent of sales and soft drinks only 18 per cent, so it can be classified as a dominant-product company. Nippon Steel's sales of iron and steel comprise about 82 per cent of the total, so it too can be classified as a dominant-product company. A related-technology company (RT) is one where one product accounts for less than 70 per cent of sales and technologically related product groups account for more than 70 per cent of total sales. The use or purpose of the products will be variable. Hitachi, Toshiba and NEC belong to this type. A related-marketing company (RM) is a company where the sales of one product make up less than 70 per cent and marketing-related products more than 70 per cent of total sales. Here again the use of purpose of the products might be diverse. For example Konica produces photographic film, cameras and, more recently, photocopiers. These products are marketing-related but the technologies are different. A related-marketing and technology company (RMT) is one where one product accounts for less than 70 per cent and marketing-related and technology-related products 70 per cent of total sales. Canon can be classified as an RMT because cameras, photocopiers and printers are mutually marketing- and technology-related and the total sales of these products account for more than 70 per cent of all sales. (The marketing-related

Use / Technology	Narrow		Wide	
			Marketing, narrow	Marketing, wide
Narrow	S : single-product company (Toyota, Nissan)	D : dominant-product company (Asahi Beer, Nippon Steel, Komatsu)	RMT : related-marketing and technology company (Matsushita, Sony, Canon)	RT : related-technology company (Hitachi, Toshiba, NEC)
Wide			RM : related-marketing company (Konica, Lion)	U : unrelated-product company (Kanebo, Asahi Chemical, Kubota)

Notes:
See main text for the statistical definitions of the categories. To find out whether two products are marketing-related or not, we used the numbers in the Standard Product Classification (SPC) index. The SPC, unlike the Standard Industry Classification, classifies products by similarity of use and marketing. If the first two digits of the SPC numbers for two products are the same, then they are marketing-related. For example the SPC number of beer is 07221 and that of soft drinks 0712, so they are marketing-related. To determine if two products are technology-related, we used the number of Standard Industry Classification, which classifies products according to their production technology. If the first two digits of the SIC numbers are the same, the two products are technology-related. For example the SIC number of fertiliser is 261 and that of plastics is 2637, so they are technology-related.

Figure 4.2 Classification of product mix

product group need not overlap exactly with the technology-related product group.) An unrelated product company (U) sells less than 70 per cent of marketing-related products and less than 70 per cent of technology-related products. For example Ube Kosan, which was originally a coal-mining company, now produces fertilisers, plastics, cement and industrial machinery.

In summary, use or purpose is the key element in product mix because product life cycle differs according to use. If the uses of the products of a company are limited, it will be a specialised company (S and D). If the uses of the products of a company are quite varied, it will be a diversified company. If the products have different and varied uses, that is, if they are marketing- or technology-related, the company will enjoy synergistic effects, giving it strong competitive power. For instance cameras and printers have different uses and life cycles but they are marketing- and technology-related.

It should be noted that the classification used is based on existing products and does not refer to changes of products. This is a broader concept than Ansoff's (1965) definition of diversification (which implies the addition of new products), and is more tightly coupled to present performance than to the addition of new products. Much research has been carried out on diversification, using such measures as the Herfindahl index (see Oster, 1994) or entropy (see Hoskisson *et al.*, 1993). While it may be easier to measure diversification by its level rather than its style, to do so neglects key intervening variables and cannot explain the cause and effect relationship between strategic decisions and performance. Strategic decision-making determines what kind of product mix is desirable by selecting the type of business the company should be in and the new products to be developed. The level of diversification is of secondary importance.

DIVERSIFICATION AND PERFORMANCE

The main objective of a product-mix strategy is to utilise growth opportunities by selecting a growth product or a product in the earliest stages of its life cycle. In order to grow, the company has to change its product mix and for this purpose the products have to be growth products – particularly in the area in which the company's share of the market is already high. If its products have similar characteristics and can use the same research capabilities, production facilities and marketing resources, then the diversified company can achieve high efficiency by using large-scale processing or synergy (Penrose, 1959; Ansoff, 1965).

A further objective of the product mix strategy is to attain stability by diversifying the use of products; in other words the performance of each product should be independent. This is emphasised in financial theory (Markowitz, 1959; Sharpe, 1970). To attain stability of performance, it the use of each product is different, production capability and marketing capability must be similar.

The relation between the diversification style and financial performance of 203 manufacturing companies for the period 1983–93 was examined. Table 4.1 presents the results. The growth rate was measured by the growth of the companies' unconsolidated sales (consolidated financial statements were not available until 1999). Profitability

Table 4.1 Types of diversification by synergy and performance, 1983–93

Types of diversification[2]	*Number of companies*[1]	*Growth rate of sales*	*Return on investment*	*Standard deviation of ROI*	*Equity ratio*	*Total performance*
Specialised (S)	33	2.28	7.40**	2.70	43.46**	10.49
Dominant (D)	45	2.46*	6.11*	2.32	38.42	12.71*
Related marketing and technology (RMT)	29	2.77*	5.01	2.97	39.37*	11.52*
Related technology (RT)	85	1.84	5.44	2.29*	34.55	10.84
Related marketing (RM)	4	0.85	4.13	1.92	36.23	8.92
Unrelated (U)	7	0.69	5.59	1.75*	24.66	9.06
Average		2.13	5.81	2.45	37.25	10.89
Standard deviation		3.58	3.31	2.09	16.61	6.29

Notes:
1. Survey of 203 manufacturing companies.
2. See main text for the statistical definitions of the categories.
3. Asterisks indicate degree of significance: * < 10%, ** < 5%; only the better performances have these marks.
4. Total performance = growth rate + ROI −½ standard deviation of ROI + $\frac{1}{8}$ equity ratio.

was measured by ROI (before interest and tax, unconsolidated). Stability was measured by the standard deviation of ROI and the equity ratio. Overall performance = growth rate plus return on investment minus 1/2 standard deviation plus 1/8 equity ratio (weight was subjectively given to each item, considering the utility of each item).

According to the survey, the best growth rates were achieved by D (dominant product) and RMT (related marketing and technology) companies. The return on investment was best among S (specialised) and D companies. The standard deviation of ROI was smallest among RT (technology related) and U (unrelated). The overall performance was best in D and RMT companies.

Performance is affected not only by the style of diversification but also by the differentiation of products and the growth rate of the industry. Table 4.2 presents the results of an analysis of these effects. We classified the 203 companies in the sample into companies with differentiated or undifferentiated products. The differentiated product group had a higher rate of return with a smaller standard deviation. This group also had the highest industry growth rate and higher performance figures. These data suggest that, first, the growth rate of sales was high among RMT and D companies. These companies could select high-growth products. Second, the return on investment was best among specialised companies, probably because the products were differentiated. Third, the standard deviation for return on investment was smaller among RT and U companies whose products had diverse uses or markets. Fourth, the differentiated products group had higher rate of return and a lower standard deviation in respect of profits.

However these findings have to be put into context. Considerable research has been undertaken on the effects of diversification but no unanimous conclusion has been reached on its effects; our findings are but one example although they may be taken as representative or indicative of the relations prevailing in Japanese manufacturing at the time of the research.

Table 4.2 The effects of product differentiation, 1983–93*

Industry	*Number of companies*	*Return on total investment (%)*	*Standard deviation of ROI (%)*	*Equity ratio (%)*	*Growth rate of industry (%)*
Differentiated:					
Dairy products	9	5.90	1.00	39.0	2.9
Condiments	10	6.99	0.94	49.0	3.0
Pharmaceuticals	35	10.06	0.55	53.3	4.3
Construction equipment	10	5.56	2.98	44.4	1.0
Domestic electrical appliances	30	6.43	5.65	50.1	2.6
(Average)		(6.99)	(2.22)	(47.20)	(2.76)
Undifferentiated:					
Synthetic fibres	6	6.29	1.85	34.1	−1.3
Paper and pulp	29	5.27	3.85	25.1	0.2
Chemicals	6	5.82	4.38	21.0	−0.6
Oil refining	11	5.11	0.85	20.0	−4.2
Cement	6	5.73	3.38	25.9	0.8
Iron and steel	54	5.28	6.73	22.6	−1.7
(Average)		(5.58)	(3.50)	(24.78)	(−1.35)

* The 203 companies in Table 4.1 are included in this larger sample.
Source: Japan Development Bank, *Financial Data Handbook* (1995).

Long-term trends

The long-term diversification trend among 102 companies was examined for the period 1962–93, the results of which are shown in Table 4.3. While there was a weak trend towards diversification, over one third of the companies had a specialised product mix. The majority of diversified companies had a technology-related (RMT or PT) product mix, which is one of the features of successful Japanese companies. RMTs showed the best performance throughout the period in question.

DIVERSIFICATION AND RESOURCE STRUCTURE

Product mix affects organisational structure. This relationship is typically assumed to be one of fit or mismatch (see Scott, 1971; Greiner, 1972; Stopford and Wells, 1972; Channon, 1973; Rumelt, 1974; Miles and Snow, 1978; for a recent analysis that encompasses much earlier research on the relation between organisational change and various aspects of organisational performance, see Donaldson, 1999). Table 4.4 presents a typical representation of the fit between strategy and structure. As in any scientific field, a number of assumptions undergird the relationships that are posited, which we shall now examine.

Top management

It is assumed that a specialised company has a simple product mix and a small number of top managers with access to sufficient strategic information. It is also assumed that a diversified company needs to have a team at the top because it requires more complicated strategic information. However we have not found this to be the case. In Japan both specialised and diversified companies have management committees. There are differences, however, in the areas of expertise of top management: technology-related companies (RMT and RT) have more engineering or natural-science graduates at the top (Table 4.5), while marketing-related companies (RM) have the smallest percentage of engineers at the top. There is no difference in the ages of directors.

Table 4.3 Transition or product-market strategy and structure, by number of companies, 1962–93

		1962	*1967*	*1972*	*1976*	*1980*	*1993*	*1993 (%)*
Product-market strategy	S[1]	18	18	16	14	14	33	16
	D	30	29	23	25	23	45	22
	Subtotal	(48)	(47)	(39)	(39)	(37)	(78)	(38)[2]
	RMT	14	15	19	21	23	29	15
	RT	25	24	28	27	30	85	42
	RM	6	6	6	6	6	4	2
	U	9	10	10	9	6	7	3
	Subtotal	(54)	(55)	(63)	(63)	(65)	(125)	(62)
	Total	102	102	102	102	102	203[3]	100
structure	Functional		55		46	43	39	19
	Functional and divisional[4]		7		11	13	26	13
	Subtotal		(62)		(57)	(56)	(65)	(32)
	Divisional		40		45	46	127	63
	Total		102		102	102	192	95
							11	5

Notes:
1. For the definitions of S, D, etc., see main text.
2. Between 1962 and 1980 the number of companies is almost the same as the percentage.
3. Includes the original 102 companies.
4. Functional for major products and divisional for minor products.

Table 4.4 Assumptions about the fit between strategy and structure

	Specialised company	*Diversified company*
Top management	One man	Team
Research	Centralised	Decentralised
Head office	Large	Small
Departmental structure	Functional	Product division

Research and advertising

Predictably, technology-related diversifiers (RMT, RT) have higher research expenditures, while marketing-related diversifiers (RMT, MRM) have higher sales promotion expenditures (Table 4.5). For example Hitachi (RT) and Sony (RMT) spend about 10 per cent of their sales receipts on research and development. Dominant product companies such as manufacturers of pharmaceutical products and machinery, and some car manufacturers, have high R&D ratios. For example Honda (D) spends about 7.8 per cent of its sales receipts on R&D.

One example of a very successful R&D-intensive firm is Canon. Canon started out as a camera manufacturer in 1933 and continued to grow for 70 years by changing its product mix and increasing its core competencies in lenses and reprography. Canon's consolidated sales in 1999, at 2900 billion yen or about $29 billion, were about ten times higher than those of Nikon, Minoruta and Asahi. Canon introduced new products by utilising its technological core competencies and increasing and enhancing these competencies. New products were introduced as old technologies became outdated and new technological breakthroughs were made. It spent more than 10 per cent of its sales receipts (unconsolidated) on research and development. These R&D costs were spread by expanding sales worldwide, so that foreign production accounted for 25 per cent of consolidated sales, and foreign sales, including exports, accounted for 85 per cent of all sales (Table 4.6).

Foreign direct investment rate

Foreign direct investment is high among specialised and technology-intensive companies (see Table 4.5 for an analysis of types of diversification). Some specialised companies such as Nissan, Toyota and Honda have very high foreign production ratios, as do some electrical

Table 4.5 Types of diversification, top management, R & D expenditure and other strategies

Types of diversification	*Number of companies*	*Ratio of technology background of directors*	*Average age of directors*	*R & D ratio over sales (%)*	*Sales promotion ratio over sales (%)*	*Overseas production ratio over consolidated sales (%)*
Specialised (S)	33	0.37	58.1	2.39	1.42	24.50**
Dominant product (D)	45	0.42	58.6	4.29	4.92	18.10*
Related marketing and technology (RMT)	29	0.41	58.0	4.29	6.16**	16.90
Related technology (RT)	85	0.49*	58.9	4.54*	1.79	15.90
Related marketing (RM)	4	0.29	58.5	2.98	11.91*	2.50
Unrelated (U)	7	0.50	59.8	4.03	2.51	9.40
Average		0.45	58.6	4.04	3.54	15.64
Standard deviation			1.93	3.05	5.86	15.76

Notes: Asterisks indicate degree of significance: * < 10%, ** < 5%; only the better performances have been marked.

Table 4.6 Transition of strategy and structure at Canon

	1950	*1960*	*1970*	*1980*	*1990*
1st year sales:					
Billion yen	0.4	4.2	44.8	240.7	900.0
10 year growth rate (%)	30.2	27.8	18.6	15.8	4.0
Products	Stills cameras Movie cameras Lenses	EE cameras Reflex cameras Calculators	Copiers Facsimiles Razor printers Word processors	Cameras (23%) Copiers (38%) Facsimiles etc. (31%) Optical products (8%) (1985)	Cameras (11%) Copiers (28%) Computer related (48%) Optical, information related (13%) (1996)
	(Cameras)	(Camera-related products)	(Image products)	(Image and information products)	(Image, information and communication products)
Market	Home market	Exporting	Foreign direct investment	Multinational enterprise	Multinational enterprise
Added core technology	Optical Fine mechanical	Electronics Physics Chemistry	Software System Material technology	(As for 1970)	Biotechnology Energy technology

Table 4.6 *Continued*

	1950	*1960*	*1970*	*1980*	*1990*
		Fine optical	Communication		
Production	One plant	Multiple plants	Foreign direct investment	Optimum production location, domestic and overseas	Optimum world-wide production location
Sales channels:					
Domestic	Outside wholesaler	Construction of direct wholesale channels	Strengthening direct sales channel	(As for 1970)	System integration Customer-oriented system
Overseas	Outside agency	Construction of direct wholesale channels			
Personnel	Recruiting	Training	Utilisation of human resources	Enhancement of creativity	Enhancement of entrepreneurial spirit
Organisational structure	Functional	Preparation of product division	Product divisions Functional committees	Group divisions	Global management system

Table 4.6 *Continued*

	1950	*1960*	*1970*	*1980*	*1990*
Long-range planning		1st long range plan (construction of resource structure for diversification)	2nd long range plan (1968–72) (diversification and expansion of production capacity and sales channels)	Excellent company plan (1976–81) 1000 billion plan (1979–1989) + medium-range plan)	Global company plan (1988–92)
			3rd long range plan (1973–77) (to image industry, to knowledge-intensive products. Development of multinational management)	2nd excellent company plan (1982–86) (strengthening Canon group, creative new product development capability, strengthening resource structure)	(to global company and information industry)

Source: Corporate Planning Office, Canon Inc.

products companies, for example Kenwood (which manufactures domestic appliances), Tiac (PC-related parts), Yuasa (batteries) and Bridgestone Tires. However, because of the large variations we cannot generalise that specialisation is related to a high ratio of foreign production, but we can say that high research expenditure is related to a high foreign production ratio.

Organisational structure

The use of a divisonal structure has increased (see Table 4.3) but it should be noted that the product division structure in Japanese corporations is rather hybrid in style. Divisions do not have research laboratories or a marketing department, and they are under strong head office control, which has a sizeable specialised service staff. In other words the head office is strong and many divisions exist as a hybrid of the functional and product division structure (this will be discussed further Chapter 9).

NEW PRODUCT DEVELOPMENT

New product development and performance

There are a number of well-known Japanese examples of market dominance based on new product innovation. Seiko was the first company to develop the crystal quartz watch, with which it carved out the largest share of the world wristwatch market. In the 1950s Honda developed a small but high-quality motorcycle after carefully studying the European models that raced in the Isle of Man TT races, and it soon dominated the world market. After the Second World War, German lens expertise was locked up behind the iron curtain in East Germany. While innovations in European cameras stalled, Japanese camera manufacturers were sensitive to changing demands and changing technology. They developed the electronically controlled automatic shutter mechanism, the compact camera, the autofocus camera and the digital camera and successfully displaced the previously dominant German camera manufacturers.

A new product may involve the creation of a new brand, which is added to the company's existing products, or it may be the result of an improvement that can extend the life cycle of existing products (Table 4.7). The reason why use is made the basis of classification is that when

Table 4.7 Classification of new products

Technology	*Same or similar use as the existing products*	*Different use*	
		Marketing related	*Marketing unrelated*
Same or similar as the existing products	(H) Existing product improvement, complementary products	(A) Marketing-and technology-related new products	(B) Technology-related new products
Different technology	(J) Substitutes, complementary products	(C) Marketing-related new products	(D) Unrelated new products
Example of new products of a camera and film company:			
Same or similar technology as the existing products	(H) (Film and cameras) ASA 400 colour film	(A) Floppy disks	(B) Industrial chemicals
Different	(J) Video cameras, digital cameras	(C) Copying machines	(D) Housing construction

Note: If companies are diversified, most new products are classified as (H) or (J).

use and purpose differ, so does the product life cycle (for a similar classification see Ansoff, 1965; Pessemier, 1977). An example of new product classification when the company's main products are cameras and film is shown in Table 4.7.

The classification in Table 4.7 is similar to the classification of product mix shown in Figure 4.2 – the same principle is used. However, while product mix is a static snapshot of what is marketed at a particular moment, new-product development is a process that involves changes to the product mix over time. A defect of this classification is that if the company is already diversified, then all the new products fall into the (H) or (J) classification in Table 4.7.

Product development model

Development strategy

New product development starts with goal setting (Figure 4.3). Basic policies on new product development are decided by means of a long-range strategy for a desirable product mix. This strategy is the result of research and planning. Many Japanese corporations use long-term planning to decide the future domain of the firm and the core competencies that will be required. Having a clear goal-oriented policy makes it easier to evaluate new ideas when they emerge: they are judged not merely in terms of their innovativeness but also in terms of how they fit in with the overall strategy. While it would be incorrect to say that all new product ideas result from long-term planning, it is the case that the majority of Japanese corporations use long-term planning, to construct policy on their future domain of operation (long-term planning is studied in Chapter 8).

Information collection and concept creation

Developing concepts for new products means generating new ideas. These underline the unique sales proposition being promoted for a new product and may determine basic policy towards it. But new ideas alone are insufficient: information is required on both emergent and projected demand as well as new technologies. Where do original new ideas come from? Our survey suggests that, not surprisingly, the head office and development departments are important sources. As one might expect corporate rather than operational departments are the most important sources of innovation.

Experimentation and testing

Many problems are encountered at the development stage, requiring repeated technical experiments until a prototype is produced and a method of mass production is found. It is at this stage that concurrent engineering is carried out. The new product development process is not one-way and includes many feedback processes. The evaluation of experiments provides feedback to the concept-developers and acts as the basis of the second round of experiments. These are tried on the new model, which is then evaluated and the findings act as feedback to the original concept. This feedback process is made possible by the concurrent development system (Table 4.8).

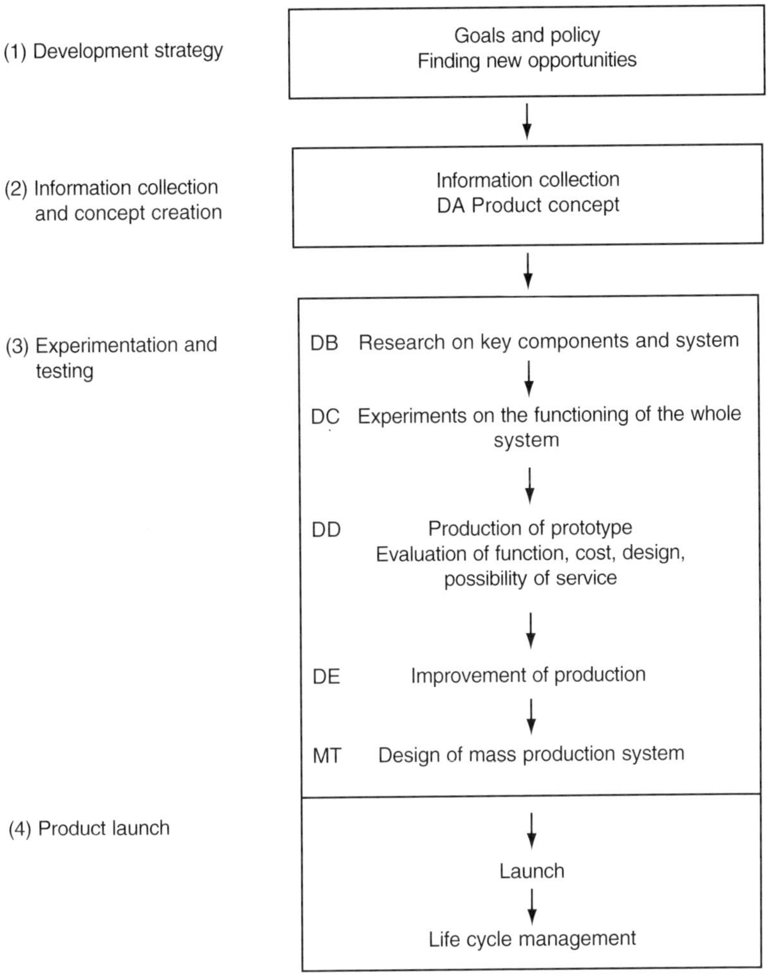

Notes:
1. DA and other symbols are used at Canon (Yamanouchi, 1989)
2. In D processes many ideas are presented and experimented with. There are many feedback and spiral processes. In the case of marketing-intensive products such as cosmetics and food, the process is almost the same: for processes such as concept-research on the subsystem, total integration will be used. Many alternatives with regard to scent and packaging are tested by inside and outside panels.

Figure 4.3 New product development model: technology-intensive product

Table 4.8 Concurrent engineering and sequential engineering

	Concurrent engineering (soccer style)	*Sequential engineering (relay race style)*
Relation with suppliers	In alliance, participation from development stage	Open bidding
Interface between development, production and marketing	Project team of members from development, production and marketing. Parallel development by three departments	Development → production → marketing
Number of alternatives	Many alternatives in the early stage, then reduced to best one.	Few alternatives
	Frequent feedback	One-way process
Development period	Short	Long
Problems	When an earlier stage in design is changed, later-stage work has to be done over again	Longer development time and uneconomical designing

Source: Adapted from Kurokawa (1997).

Product launch

The new product is launched after trial sales to evaluate the market. If the decision is made to continue, then after-sales service activity starts, which also provides valuable feedback on the product in use.

Key success factors

Support of top management

Successful new product development (Figure 4.4) needs the support of a top management with a positive attitude towards the development of new policies and new directions. When a Sony project team developed the Walkman, five million of which are sold each year throughout the world, the marketing department opposed the decision to proceed. However the president of Sony overruled them – he was clearly closer in tune with customers than were those who were supposed to be, a story that will be recounted later.

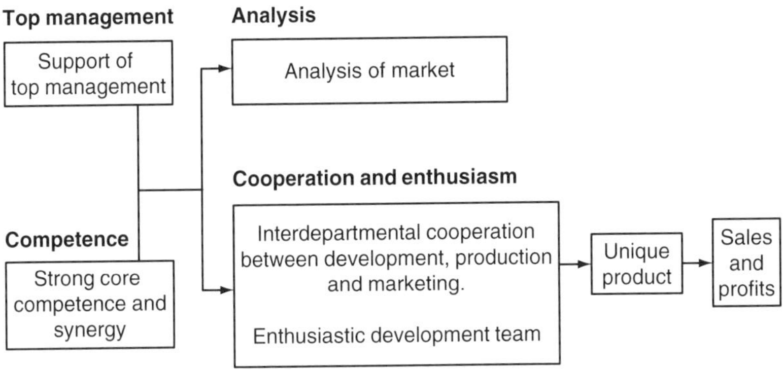

Figure 4.4 Success factors in new product development

The reasons why the support of top management is needed are as follows. First, new products do not immediately produce profits – in fact they may eat into the current profits – so it is necessary to allocate resources by ascertaining future needs. Second, new product development is risky and has a success rate of only 20–30 per cent. Companies know that the majority of projects will be discontinued or abandoned. Third, product divisions do not like to use their resources to bring on new products because this reduces their profit level, and if the new products are successful, this may erode the prestige of the existing products, around which the identity of the division is built.

When the direction offered by top management is not based on appropriate information it can result in failure. The president of one Japanese textile company was an outstanding leader in his early career but later became a tyrant with a firm belief in the omniscience of his own judgements. This caused a number of business failures in unrelated businesses, such as oil drilling, video players and cosmetics.

Strong core competence

New product developments requires high-level capabilities for development, production and marketing, which need to be supported by synergistic relationships with the present products. Successful new products reinforce existing core competencies, as was the case with Canon.

Market analysis

The Walkman was made possible by good market analysis, in addition to technological developments. There was a trend for stereo players to become smaller as a result of technological innovations in miniaturisation, the sale of these sets was increasing and the tape as a medium of music was becoming more popular than records. The technology of miniaturisation combined with the concept of portability led to the Walkman, which took three years to develop after the idea was first put forward. During the course of development the response from sales channels was very cold. It was thought that an expensive small player without a recording capacity would be unable to compete with the small taperecorders that were flooding the market, so the development team was not confident of success. But when the president of Sony inspected the prototype Walkman he declared that the sound was splendid and encouraged the development team to go ahead, thus giving a great stimulus to the team and the project as a whole. So not only was information on market trends important for success but also support from the top played a great part in encouraging continued development.

Another the story concerns the development of the Toyota Lexus. In the mid 1980s Toyota decided to develop a flagship car to commemorate the fiftieth anniversary of its founding. Designers, engineers, product planners and product managers got together to form a working group and brainstormed ideas for the flagship car. They had studied the specifications of the Mercedes Benz and considered the feasibility of designing the new model by benchmarking the performance of the Mercedes.

In 1985 the study team went to the US to research the requirements of the luxury car market. (Toyota thought that the new car should be mainly exported to the US). The intention was to get around the trade friction that existed at that time between the two countries. As a result of an earlier voluntary agreement Toyota could export only a fixed number of cars to the US, so ideally these should be profitable cars with high value added. It was for this reason that Toyota decided that the Lexus would be a luxury car. Between May and August 1985 the study team went to Chicago, Los Angeles, Miami and Denver to research how a luxury car aimed at the US market should be designed. They also visited Toyota dealers and other dealers of imported cars to gather marketing intelligence. Toyota used focus group interviews (which in 1986 was a new technique) to develop the design concept.

Toyota gathered information on the purchasers of other luxury cars (such as the Audi 5000, BMW 528e, Mercedes 190E and Volvo 740 and 760) in New York and Los Angeles. Then potential consumers were divided into two groups and the Toyota study group asked them to describe their idea of an ideal European, American and Japanese car. The study group also investigated the reasons for the purchase or non-purchase of each brand of luxury car. Such interviews were carried out four times in all. Five designers in the study group stayed at Calty (Toyota's design centre in California, employing 42 staff) for two months to observe the life style of luxury car owners. They identified two types: people aged 50 and over, often blue-collar retirees, who seemed to prefer traditional American luxury cars such as the Cadillac; and typical yuppies, such as doctors or lawyers, who preferred an imported car such as a Mercedes or BMW. Toyota chose the latter group as the target market for the Lexus because the Cadillac was seen as representing lower quality and somewhat traditional rather than contemporary values. Eventually Toyota decided that a luxury car should be defined in terms of status and prestige, high quality, high resale value (investment value), high performance (handling and ride) and safety. Calty proposed three models and developed a 1:5 scale model for each. Meanwhile the design centre in Japan proposed a number of models, three of which were developed into full models. These were then appraised in a design clinic conducted by about 10 Toyota employees, including people from the sales division (Kurokawa, 1997).

Merely collecting information does not produce new ideas automatically: a change in existing ideas is needed. For instance people in Alaska who think they do not need refrigerators because of the cold might be persuaded that there is a need to prevent foods from freezing and that refrigerators are the appropriate technology for this. Many other successes are achieved through 'reversing' information. For instance it is difficult to sell pianos to families living in apartments so Yamaha developed a 'silent piano' – an ordinary piano with optional headphones whose use eliminated the audible output to others in the apartment or building. Another example is the success of small retail stores at the neighbourhood level. Seven Eleven Japan has more than 7000 stores that are more conveniently located than the large retail outlets. In this case convenience of location was the unique selling point. Many people have cameras but often forget to take them with them or do something on impulse that they want to record – thus the disposable camera was developed by Fiji Film. It is well known that metals cannot hold a constant temperature because they conduct heat.

However, creating a vacuum between two metals led to the development of the metal thermostat by Nihon Sanso.

Concurrent engineering and creativity

Concurrent engineering is made possible by cooperation between departments. As will be explained, concurrent engineering not only shortens development time but also increases the range of alternative designs for new products.

The successful development of new products requires a considerable effort because it entails the solution of mutually contradictory problems, such as creating something that is very small but also of very high quality. Non-routine environments work well for creative innovation. Japanese researchers in laboratories are considered to be one of the hardest working groups in Japanese firms, often returning home at midnight or even sleeping in a bed at the laboratory. They are motivated by a number of factors. Not only the job itself and the satisfaction to be gained by intellectual and practical problem solving, but also top management recognition, the pride of the team and the fact that there are prizes for success (Kono, 1987c). However, unlike in Silicon Valley, stock options are rarely given to development team members in Japan.

Reverse thinking in new product innovation

The following are examples of successful new products borne out of reverse thinking that challenges common sense. Matsushita has developed a bulb-style fluorescent light. Unlike the usual rod type it can be used anywhere where an ordinary electric bulb is used but it is more efficient and has a longer life span. It costs a little more but offers far greater utility than the standard bulb. The hybrid car developed by Toyota similarly enhanced the limited utility that previous electric vehicles had offered. The Prius is a self-rechargeable electric car with a back-up petrol engine that can be used to power the car when its load is heavy and to recharge the battery. Sometimes demand is hardly foreseeable: who could have guessed that the Tamagochi (literally 'egg-watch'), developed by Bandai, would immediately sell millions when it was introduced? Tamagochis are small, battery-powered toys that display an image of a rabbit or some other animal, or even a human infant. The toy has to be attended to almost constantly or the animated animal or child will 'misbehave' or 'die' (but it can be revived) – just like a real pet but without any of the inconvenience and mess that a live animal entails, especially in a small apartment.

CONTEMPORARY TRENDS

The business environment in Japan is changing in a number of ways: it is becoming more globally competitive, environmentally sensitive, and subject to new opportunities and preferences that may make older ways of doing business less relevant (Figure 4.5).

Mega-competition

For some time Japanese companies have been able to enter foreign markets and establish production plants, for instance Japanese inward investment in Europe dates back to the penetration of the UK market in the 1980s. Now that competition is becoming more global, foreign companies are able to enter the Japanese domestic market and have met with success in computers, financial services and cars, something that was previously unimaginable. In the past it was difficult for foreign companies to gain a foothold because of government regulations, but these are now being relaxed.

The rules of the business game are changing, for example Microsoft Windows 95 eroded the strong position of previously dominant

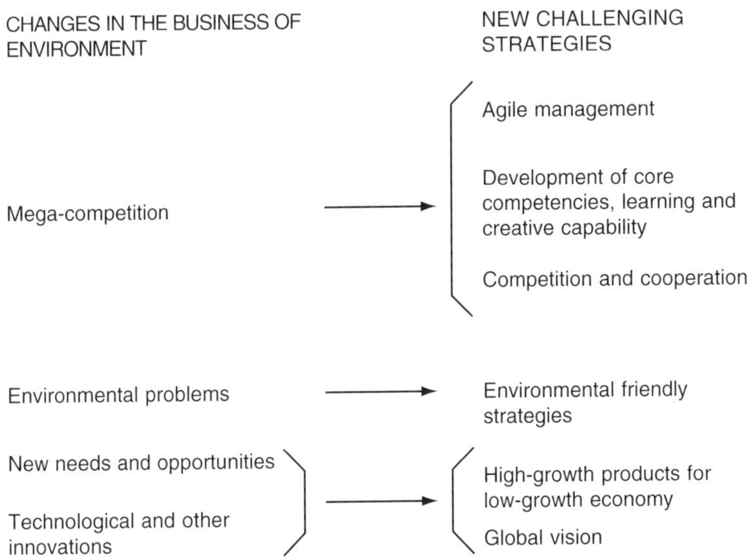

Figure 4.5 New strategic issues

manufacturers such as NEC in the personal computer market by becoming the industry standard. Technological innovations are also lowering entry barriers; for example firms previously confined to the chemicals industry are moving into pharmaceuticals. In order to cope with these new competitive conditions firms require agile management, strong core competencies and the ability to forge cooperative alliances for enhanced competition.

Environmental problems

Before a new product is developed, firms have to consider its impact on the environment by carrying out a life-cycle assessment. From the design stage firms have to consider how to reduce energy consumption and how to recycle components. Environmental as well as market design considerations now have to be factored into the design process.

New needs and opportunities

Even during the difficult economic times in the 1990s, new needs and opportunities were emerging. The sale of mobile phones and personal computers increased markedly, as did the sale of Canon printers and the industrial demand for the small ball bearings manufactured by Minebkea (Nihon miniature bearings). In the retail, sector, a mature sector with seemingly little opportunity for growth, Seven Eleven Japan grew rapidly by carefully selecting locations and product mix, using information technology as well as employing the services of consultants. Some of the companies that have been able to seize these new opportunities have grown into big businesses with sales of over $10 billion, for example Sharp, Canon and Seven Eleven Japan.

Innovative trends in new product development and divestment

Agile management and customer orientation

As the life cycle of existing products becomes shorter their speed of development accelerates so that new models appear almost every six months. The many examples from the 1990s include new types of camera (such as smaller automatic cameras and digital cameras), toilets (such as the Washlet, a Japanese 'superloo'), PCs (miniaturised personal computers such as that pioneered by Sony), recreational utility vehicles (such as the Honda CRX) and global positioning car navigation systems. Thus speedy development is becoming more important. A number of factors aid speedy development, such as the use of computers, supply chain management and concurrent engineering.

One example of rapid innovation is that by the women's clothes manufacturers Five Fox. Unlike other clothing manufacturers, Five Fox (with 4000 employees, sales of 100 billion yen and a profit of two billion yen in 1999) is a high-growth company. The design team does not simply import foreign ideas from Milan, New York or Paris but develops some original concepts. (However its best seller in recent years has been 'boyish' black suits for women, which would seem to tap a universal rather than a specifically Japanese style choice). It has an agile management team leading an equally strong and very successful design team, who have been able to forecast future trends and lead the way in ready-to-wear fashion. The management and design teams are partly motivated by the fact that their annual salaries are linked to profits. By using the supply chain system the company is able to increase production quickly once designs find favour with consumers. The company requires its subcontractors to operate 24 hours a day if necessary to enable additional orders to be manufactured in 10 days. Production before any given season is only 10 per cent but by the end of the season 75 per cent of the garments have been sold at the list price. To understand the competitive edge here, one must understand that for other manufacturers such sales usually amount to only 60 per cent. Other companies produce garments before the season in anticipation of demand. Because of the reputation of Five Fox products, the high turnover and the high ratio of sales per square metre of retail space, first-class department stores are more than willing to rent space to the company to establish its own in-store boutiques.

Core competencies

Successful new products are based on corporate core competencies, which are in turn reinforced by the success of these products. Core competencies accumulate and change, as we saw in the case of Canon. One company that has developed a unique core competence is a book publisher – seemingly a very traditional field.

The publisher Benesse Corporation was established in Okayama in 1955 and entered into the educational book market. In 1998 its sales amounted to 190 billion yen and its net profit was 14 billion yen. Today it employs about 2000 people. The company fosters close customer relations and provides one-to-one service through market segmentation, but the production and mailing of educational materials are centralised at its mass production and distribution facilities. It has continuously expanded its market – from high school students to young children, then to their mothers and other adults, and then to a market specifically for older people. The company does not sell its products through bookshops but through a one-to-one sales network. The company determines what kind of textbook is being used at a school and then mails schoolchildren an example of a study guide, with specific test questions to be answered. To subscribe, one must pay a six-month subscription in advance, the cost being 5000 yen per person per month. The children's answers to the questions are sent to 'red pen teachers', who mark their papers and send them private letters of guidance and praise, using cartoons to make their points. These teachers are usually retired women with teaching experience. Thus a one-to-one personal network is constructed that combines extracurricular coaching based on highly specific tutoring with a captive market for the books published by the company. Recruitment starts as early as possible. For infants, videotapes are distributed to teach language and the concept of numbers.

The original material for training is mass-produced in Okayama, where the head office is located, and distributed by the in-house post office. There is an automated plant that prints materials, packs them and addresses the envelopes, as well as a computer-based mailing system. The core competence of Benesse resides in its network marketing and individualized one-to-one service. It has succeeded in attracting over four million subscribers, all of whom pay cash in advance. Hence there is a high cash-flow basis to the business. The company projects a clear image that it is selling 'progress' to people – especially ambitious parents who do not want their children to receive any less coaching than others in their cohort.

Benesse has a research laboratory where developmental psychology is studied and a team produces textbooks and questions with the help of outside teacher-consultants. Information on student needs is collected through interviews, direct mail surveys and telephone surveys. The company has a network of 20 000 red pen teachers, offering two-way communication and long-term monitoring of student progress. A variety of teaching materials are distributed. For primary school children, enjoyable videotapes are provided to help stimulate their interest in learning, which is reinforced by setting them easy problems and offering them encouragement through the red pen teachers. Hundreds of texts are provided to middle-school children, while the marketing pitch aimed at high school students is that the resources provided by Benesse help to minimise anxiety.

Because of the importance of core competencies, many companies have discontinued products that relate least to these competencies, for example Nissan sold its fork-lift business, Toho Rayon sold its carbon fibre business, Yamaha discontinued its production of tennis rackets and skis, and in 1998 the Sezon Group (based on department stores and supermarkets) sold its chain of Intercontinental Hotels (a global chain of 150 first-class hotels) for $2.95 billion dollars.

Strategic alliances

Strategic alliances can reinforce core competencies. For instance Nintendo is cooperating with Matsushita to develop a new game machine, Sony cooperated with Toshiba in the development of its new PlayStation (Toshiba produced the memory for the PlayStation) and Hitachi and Fuji Electric have formed an alliance to develop a power semiconductor for large transformers.

Strategic alliances may also be used to aid development and produce economies of scale in production. NKK entered into an alliance with Tetint (of Argentine) to produce and market a seamless pipe, while Mitsui Ship Building is cooperating with Kawasaki Heavy Industries in the design and construction of ships. The factors contributing to the success of alliances will be studied in the next chapter.

Environmentally friendly products

The Fuji Film Company developed a single-use camera known as *Utsurundesu* or 'Quick Snap' in 1986 and launched it on the US market in 1987: 22 million cameras had been sold globally by 1992. The product had six major attributes: it had a low risk of failure, it fitted into the pocket, it was highly portable, it was cheap (under $10), it was high quality, and because it could only be used once it required no maintenance. By delivering all these attributes, Fuji provided consumers with added value. People could take quality pictures anywhere without worrying about an expensive 35 mm camera.

Fuji's 1986 launch in Japan provided US-based Kodak with the lead-time necessary to develop its own single-use camera. Using a crash programme and a cross-functional product development team, Kodak was able to achieve the fastest product development in company history. The end result was a low-cost, high-quality product with few manufacturing problems.

However American consumers were slow to accept single-use cameras and environmentalists condemned them for adding to the waste problem. Kodak responded by changing the name of the camera from ‘Fling’, indicating disposability, to ‘FunSaver’, which was ecologically less wasteful sounding. In addition both Kodak and Fuji began a recycling programme to alleviate worries about the environmental impact of their cameras. By responding to environmental concerns they reduced a major consumer barrier and US sales gradually began to grow. In addition Kodak’s strong brand recognition, built up over 100 years, helped to alleviate early consumer scepticism and in effect enlarged the market (for details see Thomas, 1995). Two implications stand out from this case: that a new product has to be tested for its impact on the environment, and that late-entry competitiors can expand the total market for a new product.

It is now normal for all new products to be assessed in terms of their impact on the environment and for life-cycle assessments to be conducted. Dai Nippon Insatsu (a printing company) sells packaging materials. These used to be made from plastic but are now made from paper so that they will not produce air-polluting substances such as dioxin when burnt. Of course, the use of paper also has environmental implications, but in principle it is a renewable resource and its toxicity is far less.

The recycling of consumer durables requires significant investment in equipment for decomposition. The cost of reusing components and materials can be higher than the savings achieved by their reuse, so Matsushita and Toshiba as well as Sharp and Hitachi are cooperating in the recycling of products. Although there are quantifiable costs associated with recycling and environmental sustainability, companies that do not consider the impact of their products on the environment face potentially far more onerous although less quantifiable costs from consumers and environmental activists. A number of companies have had to pay considerable compensation to victims of their products or processes. These cases have been well published in the Japanese financial and business press, and the demonstration effect of such settlements has served as one of the strongest reinforcers of moral behaviour by Japanese corporations. By themselves they will not alleviate ecologically injudicious activities but they are an important factor in shaping the green strategies of business (Orsatto and Clegg, 1999).

Growth products during the recession

Successful companies were able to launch growth products even during the economic downturn in the 1990s. Sony had been producing the hardware for Nintendo's game machine but Nintendo decided to change its alliance partner from Sony to Philips, where upon Sony resolved to enter into the game machine business in its own right. This was not a sudden leap into the dark because its alliance with Nintendo had enabled it to expand its technical knowledge. It had made a number of innovations to the game machine, including changing the recording medium from tape to CD (Sony had already been successful in the music recording business and so it was familiar with CD technology). It had also changed the contract system so that the risk formerly assumed by the software suppliers had been taken over by Sony. Furthermore it had acquired skills that enabled it to enter the personal computer market, such as miniaturisation skills, which resulted in its notebook computers setting the industry standard for thinness. Sony introduced its PlayStation in 1994, and in 1999 its global sales of about $6 billion contributed about $1 billion to Sony's profits. The successful new product was based on Sony's core competence in miniaturisation.

Global standards

In the past, Japanese corporations have supplied the world with many global standards, such as those embedded in cars, electrical home appliances, home video recorders and 3.5 inch floppy disks. These standards were the result of distinct Japanese management principles in practice: quality circles supporting high production standards, concurrent engineering, and just-in-time delivery systems. However global standards originating in the US and in Europe have increased in number. For instance Microsoft has set a de facto standard for the operating system of computers while ISO 9000 and ISO 14000 are European-based standards. Sony has learnt the hard way about the importance of new products following a standard if they are to be successful. Its more sophisticated Beta video technology lost out to the VHS format because the latter was freely available as a standard for other manufacturers to copy while Sony sought to control its proprietary knowledge through exclusive licence. The license strategy deterred content providers and restricted the range of contents available as rival 'me-too' manufacturers used the freely available VHS standard. Today, Sony considers the global market whenever it introduces new products and it struggles to establish the de facto standard. For instance it has been striving to unify the standard format for DVD (Digital Video Disc) technology with Toshiba in order to increase the market share for them both. If they have different DVD formats the market share for DVD technology will not expand as much as it would have done with a shared standard.

The recent trend is to try to establish the de facto standard by releasing one's patent to all-comers either with a licensing charge or, increasingly more often, without charge. The establishment of a world format for mobile phones is an example. Due to its first mover advantage Nokia, a Finnish company, set the European standard and American and Japanese corporations then tried to ensure that their technology conformed to that standard.

PROBLEMS WITH THE JAPANESE SYSTEM FROM AN EXTERNAL PERSPECTIVE

In this section we address some of the reservations frequently aired by foreign business people and academic experts when commenting on Japan. We first present some standard representations of Japanese enterprises, and then discuss the extent to which elements of the paradigm may be changing.

Growth orientation at the expense of profitability?

Standard representation. Companies rarely compute NPV (net present value) or rate of return by DCF (discounted cash flow) methods, sales estimates being seen as more important. Nippon Steel, Kobe Steel and other firms entered the semiconductor business because it was a growth area and was related to material engineering. However they suffered a loss because their sales estimates were never realised. Likewise in 1999 more than seven companies were producing digital cameras but not all of them made a profit. There are many other examples of profit estimates being neglected in new product development decisions.

The rate of return on investment for Japanese corporations is much lower than that for American corporations (we addressed this problem in Chapter 1). Foreign shareholders have significant stock-holdings in 10 per cent of all companies listed on the Tokyo Stock Exchange (more than 30 per cent of stock). These investors are unlikely to be as silent about the rate of return on investment as Japanese banks have been in the past.

Discussion. Multiple goals rather than an obsession with the bottom line remain important to Japanese companies – in the past their actions were guided by the belief that too much emphasis on profitability tended to dampen innovation. Events in Silicon Valley, where tens of thousands of venture businesses support the growth of the US economy, would seem to support some aspects of this view. Many of these firms started up without precise profit estimates. We should recognise that risk is a feature of successful companies as well as unsuccessful ones.

Imitation and incrementalism?

Standard representation. Most new Japanese products come from foreign patents. While Japanese corporations are the world's largest producers of nylon, polyester, transistor radios, TVs and semiconductors, most of these products depended on patents imported from the US and elsewhere.

Discussion. Recently Japanese corporations have produced many unique products based on Japanese patents (such as quartz crystal wristwatches, small automatic cameras, video cameras with LCDs, DVDs and Nintendo game machines). Japanese corporations, especially giants such as Sony, Hitachi, Toshiba and NEC, spend large amounts on research and development (see Table 4.5). Consequently, the number of patents lodged with the national patent office is the largest in the world. In 1996, 215 100 patents were granted in Japan (Japanese Patent Office), and in 1995 the proportion of patents registered in the US amounted to 21.5 per cent, compared with the 55 per cent share held by US corporations. (Japanese Patent Office; Asahi Newspaper, 1999, p. 254). Japanese corporations produce and sell many technology-intensive, high-value-added goods on the world market.

Lack of originality?

Standard representation. Japanese corporations tend to follow a 'me-too' strategy, consequently many companies have a similar line of products. They do not try to find unique niches.

Discussion. Because of this me-too strategy there has been severe competition in the domestic market, but this has helped Japanese corporations to attain worldwide competitive strength – that is, if they could survive such competition in the home market, then the global market seemed hardly likely to present a greater challenger. But the strong yen has changed the situation immeasurably: highly successful car companies have passed into non-Japanese hands, for example Nissan is now owned by Renault and Mitsubishi is now controlled by Daimler-Chrysler. Changes in corporate governance, employment and social contracts are inevitable. Japan is increasingly being integrated into the global economy on a new set of terms dictated by national policy failures in macroeconomic management, which should ultimately serve to encourage enterprises to be even more innovative and competitive.

Slow rate of change of product mix?

Standard representation. Japanese corporations introduce growth products to change the product mix over the long term. Consequently they have a slow rate of product-mix change.

Discussion. The case of Canon in Table 4.6 is typical. If we take the electronics products industry, the product mix changed significantly during the ten years between 1985 and 1995. Electrical domestic appliances declined from 20 per cent to 10 per cent of output, while electronic components increased from 35 per cent to 40 per cent and industrial electronic machines rose from 45 per cent to 50 per cent (Industry Analysis, 1997, Keirin-shobo).

When Japanese corporations change products they retrain their employees in the required new skills, instead of hiring and firing. Consider the case of NEC's Tochigi plant, which produced made-to-order industrial machines and medical equipment at the rate of about 10 machines a month. It was decided to change production to lithium batteries, produced at the rate of about 100 000 a month. The methods of production control and the equipment and skills needed for production were completely different. In 1997, 280 employees were asked to choose between retirement, transfer or retraining: 110 retired, 30 were transferred to other plants and 140 applied for retraining. The latter were sent to a plant at Toyama and were trained for one year, living a single life, although all were married. Knowledge of software to operate the control mechanisms of electrical engineering was replaced by knowledge of chemical materials and mass production control. The restructured plant is now being operated successfully with these retrained employees. NEC paid one year's salary plus a bonus, living expenses and transportation costs, including two home visits. Hence the cost of retraining was considerable, but by 1998 the new plant was producing 600 000 batteries a month. Firms have to spend a large amount to carry out restructuring and protect existing employment. It may be faster to hire and fire or use acquisition and divestment to change the product mix, but retraining remains the usual approach of Japanese corporations. As this example demonstrates, it often entails a sacrifice on the part of employees as well – but at least they do not suffer the 'corrosion of character' identified by Sennett (1999) as a corollary of the new capitalism in the US. (By 'corrosion of character' he means the unceremonious dumping of loyal employees who can be replaced by cheaper employees brought in from elsewhere.)

Japanese firms rarely use domestic takeovers to bring about diversification, but in foreign markets Japanese corporations aggressively acquire other companies. For example Sumitomo Rubber acquired Dunlop and Bridgestone Tire acquired Firestone. Bridgestone faced a major task in modernising the plants that were acquired and improving quality and reducing costs. Eventually, however, by dint of these acquisitions Bridgestone became the largest tyre manufacturer in the world.

Too few venture businesses being born?

Standard representation. It is estimated that new venture businesses in the US are creating millions of new jobs, but in Japan the number of new venture businesses is small. The main reason for this is that university graduates do not want to enter small businesses because it would be difficult for them to find a new job if the business failed. Under the lifetime employment system it is disadvantageous for people to change employers: they tend to be regarded with suspicion if they are not 'company men'. Large corporations develop most new ventures with the help of formal project teams (for example Seiko's quartz crystal watch, Sharp's LCD video camera and Nikon's LSI manufacturing machine). Individuals do not initiate many intercompany ventures.

Discussion. The conditions for a higher start up rate for independent venture businesses might develop in the future if there is better co-operation between universities and venture businesses. While it is correct to say that at present Japan lags behind the US in developing 'new economy' businesses, it would be incorrect to assume that it is totally out of the game, confined only to old-economy products. For instance, as Cornell (2000, pp. 14–15) recounts, in west Tokyo a new building called QFront has just opened. Two of its floors are devoted to Bit Valley (the centre for Japan's on-line entrepreneurs), and the people who are spearheading the launch of those e-business floats are now showing on the broad Topix market index. In 1999 new-tech firms dominated the 102 public offerings that were floated. Many of these are based in and around the telecommunications industry, in which were concentrated 32 per cent of the mergers and acquisitions that occurred in Japan in 1999 (to a value of nearly 2.5 trillion yen).

For the new economy really to take off, graduates will have to be employed less on the basis of where they went to university and more on the basis of the aptitudes, skills and competencies they developed whilst there. More local venture capital will need to emerge to service new stock markets such as NASDAQ, while stock option systems will have to be introduced to appeal to the new generation of graduates who are responding favourably to the rewards-based pay systems being introduced by some of the bigger companies. These may well help to create a different climate of employment and business.

The Japan of the past, of the old economy, kept returns on equity low and the turnover of bank finance high and long term, and it was this that fostered the long-term planning that made corporate innovation the norm. Today, as mergers and acquisitions accelerate in a most un-Japanese manner (340 per cent in 1999) the national Economic and Planning Agency is encouraging risk-taking and a shift towards a knowledge-based society. The foundations of such an economy have already been laid by the Japanese corporate sector. The record of the latter's postwar success indicates that it could not have been otherwise. The corporations have always driven the knowledge base: what is new in the mix is an increased emphasis on risk-taking.

SUMMARY

The product-market strategy of Japanese corporations has been to operate in the most profitable domain. Such a strategy affects performance more than the structure and operation of the firm. Even efficient operations do not bear fruit if products are at the end of their life cycle.

Product mix and diversification are measured either by the level of diversification or the type of product mix. We measure diversification by the mutual distance between the products, or their synergy relation, because this is related both to resource structure and to performance. The goal of diversification is to separate corporate performance from the effect of product life cycles, and thus for the corporation to surf the waves of growth products by utilising the synergy effect and core competencies. Technology-intensive specialisation and RMT (related marketing and technology) companies have a relatively high performance. Corporate performance is determined not only by the synergy relationship but also by the growth rate of the product and the differentiation of the product based on core competencies.

Technology-related diversifiers have more top managers with a natural science background and have higher rates of foreign direct investment. Diversified product companies have a product divisional structure but often this structure is imperfect, being a hybrid of the functional and product division structures. There are strong, centralised head offices and the research and marketing functions are under head office control. It is this, together with strong staff support from the head office and good interface between departments, that allows the mobilisation of resources into growth areas. The sense of profit responsibility by divisions is weak.

Key success factors in new product development are top management support, development based on core competencies, good interface between departments and the enthusiasm of development departments.

Recent trends in successful new product development are as follows:

- Agile management, as seen in short development times and frequent improvement of new products. The danger of this is that can create dead stock, but this can be avoided by the use of SCM (supply chain management), as in the case of the Five Fox Company.
- Mutually reinforcing relationships between the core competencies and new product development. Successful new products are produced by advanced core competencies, which are in turn reinforced by the successful new products, as in the case of Canon and the Benesse Corporation.
- Competition and cooperation. In order to survive mega-competition it is necessary to cooperate with competitors to develop new products.
- Environmentally friendly product development. Domestic appliance manufacturers now conduct life-cycle assessments to estimate the products' private and social cost from production through to

end-of-life disposal and reuse of the components. From the design stage, the company has to consider how to reuse the components, what materials can be reused, and how to reduce the products' energy consumption.
- New growth products can be found even in today's slow-growth economy. These new products have to follow global standards.

There are four features of new product development by Japanese corporations. First, the new product is developed internally because this serves to expand the internal labour market. Diversification by acquisition is rarely used. This means that products are technology and marketing related. Alliances in new product development have increased recently. Alliances are needed more in internal development than entry by acquisition because such development needs high-level and wide-ranging knowledge and incurs high development expenses. Second, there is a strong emphasis on quality and careful consideration of end-users. This is illustrated by the high ranking of Japanese goods in the US *Consumers Report Journal* and the high premium price that second-hand Japanese cars command in the US market. Third, concurrent engineering is used to carry out development simultaneously among project teams, component suppliers, production and marketing departments. It is this that makes it possible to develop a number of alternative models and faster development times. Fourth, enthusiasm is high among development teams.

The problems inherent in Japanese new product development may be seen from the perspective of non-Japanese researchers and businessmen, as follows:

- Neglect of profit estimates for new products and too much emphasis on projected sales, resulting in an overall rate of return that is very low compared with American and European standards.
- Development is imitative and incremental.
- Underuse of acquisition and divestment results in a slow change of product mix. Lifetime employment is one of the causes of this.
- There are few venture businesses in Japan and the 'new economy' has not developed as much as in the US.

5 Strategic Alliance and Vertical Integration

CONCEPT AND TYPES OF ALLIANCE

Alliances involve cooperation between more than two companies to achieve a common goal or obtain mutual benefit by combing the specific strengths of each company. Thus an alliance involves something more than merely transactions. While transactions are the stuff of business and organisational relationships in general, only those that are of more than fleeting duration and are built on some kind of relationship count as alliances. For instance any trade that occurs through market mechanisms at arm's length, using only market signals such as price or quality, is not an alliance. Neither is free trade conducted through the Internet, nor the non-specific forms of loose co-operation that sometimes occur between the members of a *zaibatsu* group (exchange of personnel or cross-shareholding) because there is no exchange of key resources. Several types of alliance can be distinguished, using various criteria.

Distinguishing alliances by the degree of integration

An alliance may be classified into four types, depending on the degree of integration involved, as shown in Figure 5.1.

Licensing

A licensing agreement involves a long-term contract to trade technological knowledge. The licensee not only pays a fee but also accepts contract specifications, for example in terms of product quality or restrictions on export destinations. The flow of knowledge is one way and occurs only once, and the scope of the knowledge is limited. Examples of licensing relationships include the production of drugs by licence, the production of colour televisions by Japanese electronics firms using the RCA patent, and the production of semiconductors by Taiwanese companies using patents held by Japanese companies.

The factors that are important to a firm when determining whether to enter into a licensing relationship include the following. First, the

Table 5.1 Types of alliance by the extent of integration

Type	*Cases*	*Determining factors (selection criteria)*
1. Market	Trade of standard products	Free competition works better
2. Loose cooperation	Interlocking stocks	Stable trade, protection against hostile takeover
3. Licence	• Production of drug by licence • Colour television • High technology product	• Licensee has related technology • Licensor wants to expand *de facto* standard • Market is not large
4. Long-term contract, without control	• Joint development of drug • Mutual OEM production	• Both partners have exchangeable technologies • Large investment for development or production • Need for early entry because of intensive competition
5. Long-term contract, one partner controls the other	• Car industry (suppliers and Toyota) • Franchise chain (Seven Eleven Japan)	• One partner is a large company • Need for vertical integration • Need to cover wide area (franchise chain)
6. Joint venture	• Joint venture (NUMMI, Fuji-Xerox) • Toshiba Silicone • Development of iron ore mine • JV in developing countries	• Both partners have key resources • Large investment for development or production • Foreign direct investment (particularly in developing countries)
7. Divestment of a department and purchase of a business	Nissan divests fork-lift truck and aerospace business	Concentration on related products

Table 5.1 *Continued*

Type	*Cases*	*Determining factors (selection criteria)*
8. Merger and acquisition	Merger of Jujo Paper and Sanyo Kokusaku Pulp	Economy of scale
9. Internal development	Almost all new product developments by Sony	For the first entry

Notes: Only 3–6 are alliances. The last three types represent a change of business within the hierarchy.

licensor must be sure that the licensee has related technologies that will enable it to produce high technology products simply by introducing limited and strictly defined additional knowledge. For instance companies in less-developed countries may not be able to produce the necessary quality of colour television, if they do not have related technologies and human resource skills. Second, the licensor might be better off selling its patents to local companies when the market in a foreign country is not large enough to warrant the establishment of a large production centre. Third, a licensor may decide to enter into a relationship as a means of creating a dominant de facto standard. Sometimes it is essential to offer patents in order to establish a de facto standard that will expand a market and create a supporting group of software suppliers. In this way leadership may be established over the whole industry. For instance in the production of DVD products, Sony was eager to standardise the format with the Toshiba group, having learnt from its experience with Beta videos.

Long-term contractual cooperation without control

Examples of long-term contractual cooperation without control include the joint research contracts sometimes established between two or more pharmaceutical companies. Often these are entered into when a drug's development costs and risks are very high. One way of buffering costs and risks and ensuring a wide ranging technological solution is to mount joint research projects. For instance Hitachi and NEC were once competitors in the production and sale of DRAM (Dynamic Random Access Memory) but recently they have begun to cooperate in the development of large-scale semi conductors, and six other companies are cooperating to fix the standard for one-giga byte semiconductors. The mutual production of OEM by two or more companies is another example.

Some of the determining factors for entering into an alliance are as follows. First, each party should have a different capability to offer. For instance in the semiconductor example, Hitachi and NEC have each accumulated relevant but different development knowledge. Second, cases where the investment cost for either R&D or production is large. Drug or semiconductor development are cases in point. Mutual OEM production is another. Third, where there is considerable competition combined with overcapacity, such as in the stainless steel industry. Stainless steel had been a growth area into which six Japanese companies, including Nippon Steel, had entered, but in the 1990s demand slowed as the recession took hold and overcapacity became a problem; thus the six companies combined to concentrate their production facilities.

Long-term contracted cooperation where one party is in control

There are many examples of such relationships in Japan, the most widespread and best-known of which are the *keiretsu* or vertical alliances that are typically seen in the car industry. For instance Nissan has Nisshokai, which is composed of 192 component manufacturing companies from whom 90 per cent of external procurements are sourced. (Nissan is expanding these sources to include Denso and Aishin, members of the Toyota *keiretsu*.) Nissan, Toyota and Mazda organise component suppliers, train them in new technologies and management systems, and encourage mutual technology learning. These techniques and practices enable assemblers to enhance the quality of cars while reducing their costs by shortening the development time through concurrent engineering, JIT, *kanban* and other systems. The system that these firms have developed is very different from that of Ford, for example, which produces 50 per cent of its components internally. The *keiretsu* system tends to breed a feeling of dependence among component manufacturers. Toyota has Kyohokai, a group of 213 component suppliers providing about 70 per cent of all components, who sell their products only (or mostly) to Toyota. Denso is one of the largest electric car-component producers in the world, selling about 45 per cent of its products to Toyota. In the past the agreement did not allow them to sell to Toyota's great rival, Nissan, a situation that changed only recently. Toyota holds the key resources, such as the overall conception, design, marketing and distribution of successful cars; thus it has controlling power. Not all component companies enter into *keiretsu*. For instance tyre manufactures and steel producers maintain mass production capacities as well as unique core competencies: they do not need to surrender their autonomy to the *keiretsu*'s vertically integrated network.

It is not only in traditional Japanese manufacturing that we find such arrangements. For instance Nintendo produces the key component in the game machine business. It also controls its software suppliers, limiting them to supplying only three products a year and making them bear the risks involved in producing the software. Other examples are to be found in franchising. Seven Eleven Japan coordinates more than 7000 retailers under its franchise system. Its head office recommends store layout and merchandising, provides a rapid and frequent distribution system, point-of-sale and other information systems, recommends new products and offers consultation by 900 store advisers. Seven Eleven holds the key knowledge, so it controls the franchisees.

Long-term cooperative contracts where one party is in control tend to exist under specific conditions. First, where there is a desire for vertical integration. Toyota requires high-quality components and to ensure consistency it does not enter into contracts based only on best price for a batch: instead it builds long-term relations with selected suppliers. Toyota provides the design parameters for the components and offers the selected suppliers a large and continuous supply of orders. Thus it is able to control its suppliers and ensure that its specifications are met. By controlling the upper stream of the value chain, Toyota is able to carry out concurrent engineering. Toyota also controls its exclusive sales channels and can sell its cars at more competitive prices because there are fewer margins to be extracted. Likewise Nintendo also needs a regular flow of attractive software for its game machines. It controls the quality of its suppliers through its monopoly of the game-machine.

Second, such contracts usually involve an alliance between a large and strong firm and a smaller and weaker firm. The controlling power originates from the unequal exchange of resources. Finally, such contracts usually occur in situations where the business activity covers a wide area. Seven Eleven Japan could not have easily established 7000 sales channels with its own resources but by bringing liquor retailers on board it was able to expand its franchise network in a short period of time.

Joint ventures

A joint venture results in a new company being established by two investing companies. Typical examples are NUMMI (in Freemont, California, established by General Motors and Toyota), Fuji-Xerox in Tokyo, and some of the coalmining and iron-ore companies established for mineral extraction in Australia by Japanese iron and steel companies. Many such joint-venture subsidiaries of Japanese companies are situated in either developing or resource-rich countries.

There are several determining factors when setting up a joint venture. First, there tend to be long-term and large-scale activities to be carried out and these can be better handled by an independent unit. In such ventures capital investment tends to be large, so risk sharing is another factor. Typically, joint ventures in mining are established for this reason. Second, companies combine to take advantage of each other's specialised knowledge. At NUMMI, Toyota could offer skills in the production of small cars and experience in participative management systems, such as *kaizen* activities, while GM could provide physical assets, skilled workers and the knowledge needed to negotiate with the unions. Third, joint ventures can often overcome political objections to multinational activities in economies where multinational penetration is a contentious issue. Developing countries often require multinational companies to establish joint ventures with local companies – this enables the local organisations to absorb techniques, knowledge and know-how from their more sophisticated foreign partners. The benefits are two-way, however: investing companies are also able to utilise local marketing expertise and political influence. Sometimes, especially in centrally managed or highly bureaucratic regimes, this may involve the effective purchase of local political support to ensure that obstacles to the venture will be minimal or absent.

Distinguishing alliances by their place in the value chain

Horizontal alliances

Horizontal alliances include alliances for research and development, mutual OEM production, the joint operation of transportation systems and the joint use of sales channels. These sorts of alliance occur where the combining of two companies' resources brings added strength (as with Fuji-Xerox and NUMMI) or offers economies of scale, for example through the joint use of a transportation system.

Vertical alliances

Vertical alliances occur between two companies in different stages of the value chain. Showa Denko (chemicals) and Asahi Glass, which market a new Freon gas substitute, agreed that Showa Denko would produce the gas and Asahi Glass would do the marketing. Hence there is no controlling relation. Yet in many vertical alliances there is one company in control, as in the case of the Toyota suppliers described earlier. Through this quasi-vertical integration Toyota can concurrently engineer and design new models and operate a JIT system for their production. By providing technical assistance to suppliers the quality of components is assured, and by concentrating orders production costs can be reduced. As already mentioned, Toyota also controls its sales channels: it has 309 dealers and 5600 sales points with 124000 salespersons, hence less profit disappears into sales margins.

The determining factors for firms to follow a quasi-vertical integration strategy include the following. First, the company being the main or key process controller of the company next in the strategic supply process. The key process holder has power over and can control supplementary firms. Second, the process should be integrated and for the important key – a 'next-to-key' process. While it is for a car manufacturer to integrate the transmission production process there is little point integrating simpler elements.

Triangle alliances

A 'triangle alliance' is where two companies cooperate to control the same component suppliers or operate the same sales channels. For instance up until 1999 Nissan and Mazda jointly owned an automatic transmission manufacturing company (the arrangement was discontinued because Nissan wanted to monopolise the key process). Likewise Toyota decided to sell GM cars in Japan through its exclusive sales channels, thus making joint use of sales channels, and Fuji-Xerox and Riko decided jointly to construct a facility to dismantle used copiers so that they could reuse the components. Another case is that of Toshiba and Matsushita, which are developing a joint transportation system and stockyards to recover used consumer electronics and transport them to recycling plants. The determining factor in the decision to enter a triangle alliance is whether scale economies can be achieved by jointly using the upper or lower section of the value chain.

Distinguishing alliances by time horizon

The nature of the cooperation between two or more companies changes over time, as shown in Figure 5.1. (The processes are similar to the conflict resolution models by Greiner, 1972, and Deutch, 1969.)

First, there may be a situation where growth delivers a win-win situation to everyone, and as a result of the companies in the alliance increasing their sales, profits and satisfaction they may seek to add depth to the alliance by developing it further. Some cases in point are those of Fuji-Xerox, NUMMI, Toyota and Denso, as outlined above. Second, some alliances are maintained only until the goals for which they were established have been achieved. Examples include the association formed in Japan for the development of semiconductors and, in the US, the association formed for the development of optical fibre by three companies under the sponsorship of NTT. In another case Matsushita Electronics engaged in a joint venture with Philips, but Philips' holdings were eventually handed over to Matsushita. Third, in other cases growth may occur despite conflict between the partners. National Thai solved the conflict between Matsushita and the Siu family over their joint venture (a case that will be explained later). Such conflicts frequently arise in alliances between Japanese firms and foreign family-owned enterprises. Fourth, sometimes the alliance is dissolved

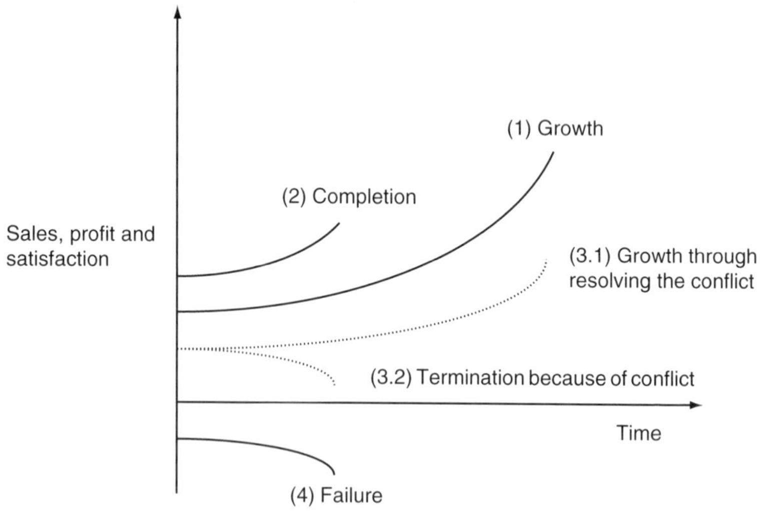

Figure 5.1 Development stages of alliances

as a result of irreconcilable conflict, in which case one of the companies usually buys out the other. For instance Hitachi acquired all the shares in GEC-Hitachi. Fifth, sometimes an alliance can end in the dissolution of the contract. Boden and Meiji Dairy Products dissolved their contract because of conflict: Boden expected more sales to be achieved through the alliance with Meiji than proved to be the case. Sixth, often alliances simply end in failure because the joint venture cannot make profit and the contract is dissolved.

Distinguishing alliances by balance of power

The balance of power between partners can be defined as equally strong, unbalanced or equally weak. Examples of equally strong partnerships are the joint venture by IHI and Sumitomo Heavy Industry to develop the LNG boat, the computer development project by Hitachi and Fujitsu (an alliance forced by MITI that did not succeed because of the rivalry between the companies), and the Fuji Film–Xerox relationship. Examples of unbalanced power are the relationships between Ford and Mazda, Seven Eleven Japan and its franchisees, Toyota and its component suppliers, and Fujitsu and Amdal. In each case the first-mentioned partner is the stronger of the two. The success or failure of alliances formed by small companies that are equally weak depends on the resources each partner can supply and their willingness to cooperate (Bleeke and Ernst, 1995), as will be studied later.

CHARACTERISTICS OF ALLIANCES

Vertical alliances

Japanese firms are characterised by numerous vertical alliances, extensive use of horizontal alliances for international operations and the widespread cross-holding of stocks to promote cooperative relations between companies, particularly those in *zaibatsu* groups. (We do not examine the latter because they are very weak alliances.) As discussed earlier, many large Japanese firms control their component suppliers and sales channels, in contrast with the US, where as many components as possible are produced by the firm itself and open bidding is used for outside purchasing, and although sales channels in the US are sometimes exclusive, car manufacturers do not control their retailers to the same extent as they do in Japan so retailers can change from brand to brand. It was for the latter reason that Japanese car manufacturers were able to easily enter the American market.

The manufacturers of domestic electrical appliances such as Matsushita, Sony, Toshiba, Sanyo, Hitachi and others also produce key components such as key integrated circuits, batteries, cathode-ray tubes and speakers. Basic materials such as steel plates, stainless steel plates and copper wire are procured from outside. Components such as cabinets, small panels and circuit boards are also bought from outside, but in this case the suppliers are organised into an alliance with (but are not controlled by) the purchasing companies through long-term contracts.

Alliances with sales channels tend to be fairly strong. Matsushita is allied with about 5000 retail stores that sell mostly Matsushita products; Toshiba has about 3000 affiliated stores. There are three types of sales outlet: exclusive sales outlets, selective sales outlets and open sales outlets. Toyota has exclusive sales outlets, although it has opened them to GM cars. Shiseido has 25 000 selective sales outlets for its cosmetics but uses open sales outlets for its toiletry range. The selective sales outlets are retailers who sell Shiseido cosmetics but are also allowed to sell other brands. Shiseido sends as many as 9000 make-up advisers to these stores to give advice to customers on make-up as well as to engage in other forms of sales promotion. Through this type of vertical integration the manufacturers are able to achieve several things at the same time: carry out effective sales promotion, maintain relatively small profit margins as the norm for the retailers and wholesalers and sustain the market price. In addition they can introduce new products to the market in a short period of time and obtain customer feedback. This also serves to hinder the entry of competitors and foreign firms into the market.

The reasons for the popularity among Japanese firms of vertical integration or vertical alliances are as follows. First, in the booming economy that existed before 1990 companies could overcome their resource limitations by using externally generated resources; also, suppliers and retailers were less independent and more group oriented so they were willing to enter into these alliances. European and American suppliers and retailers tend to be more independent and do not like to be restricted to one manufacturer.

Second, manufacturing companies can grow quickly by expanding beyond their resource limits. The technology of the large central corporations is transmitted to the small allied businesses, enhancing technological and management standards and ensuring high-quality final products. It is estimated that about 60 per cent of small manufacturing firms work under *keiretsu* arrangements, thus elevating the technology of these firms. A similar principle applies to retailers: through training and other aid from the central manufacturer, small retailers can improve their knowledge of selling and servicing high technology products. Third, vertical alliances make concurrent engineering possible. By restricting orders to a limited number of suppliers, economies of scale can be attained and knowledge accumulated.

There are some disadvantages, however. Suppliers and distribution channels in *keiretsu* or alliances are in a secure, monopolistic position, and is can lead to the neglect of innovation. For example there used to be just three producers of NTT (telephone network) telephones, all of which were black dial-phones. When Sony, Sharp and other consumer electronics manufacturers were allowed to enter the market and sold white, push-button smart-phones with memory functions the three original companies rapidly lost market share. Likewise Nissan's component suppliers used to charge Nissan higher prices than the market price.

A second disadvantage is the lack of economies of scale. If component suppliers can sell their products to several assemblers they can expand the scale of their production and achieve such economies, but if they are captive to one powerful assembler this will not happen. Nowadays Denso (the electrical component supplier in the Toyota group) also sells components to Honda and Mazda. Nissan is also reducing the number of its captive suppliers. The retailer problem in alliances is more serious. Matsushita now has 5000 alliance retailers that sell only Matsushita products, whereas formerly there were 17 000. Matsushita tries to classify its retailers according to the speciality market they serve. Salespeople (mostly the owners of the shops) cannot explain the details of all lines of high technology products such as PCs, word processors and CD players to customers, hence the specialisation. Increasingly, consumers want to compare the products of different producers and to do so they have to go to large-scale retailers or department stores. Thus both the number of retailers in captive alliances and their market share are decreasing.

In addition alliances can be inflexible compared with markets. Technology changes rapidly, and being bound by an alliance can be an obstacle to strategy change. As well as using retailers, Sony now advertises and sells goods directly to consumers through the Internet, which allows it to sell a much wider range of goods at a lower price, as well as reduce its inventory. Matsushita and Hitachi have 5000 shops in their alliances and are afraid of resistance by these shops to changes that exploit new technologies such as the Internet. By contrast there are only 2000 Sony shops and these are not totally captive – they are large shops that also sell the products of other companies.

A key feature of Japanese corporate strategy is the quasi-vertical form of integration known as *keiretsu*. *Keiretsu* have advantages and disadvantages. One advantage is that the technology of the dominating company is transmitted downstream. Disadvantages are that the protected status of the companies in the alliance can lead to neglect of learning, and costs may by higher than those determined by the market. Nissan once bought more than 90 per cent of its components from the companies in its *keiretsu* group (*Nishokai*). The *keiretsu* prices tended to be higher than the market prices, so Nissan tried to reduce the number of companies in the *keiretsu*, by buying from other companies, such as Denso, a group company of Toyota. Sometimes the managers of the component suppliers are appointed by Nissan and these personal relations tend to hinder a competitive attitude. *Keiretsu* companies may seek to sell their components to other companies: when they do, this can expand the scale of production. Other suppliers are developing module production systems. Seat suppliers now produce a complete set of seats and sell them to assemblers. This tends to increase the independence of component suppliers. Denso and Aishin Seiki have cooperated to design and produce brake systems for cars and sell this system to Toyota. The system is expected to reduce costs by 20 per cent. Toyota's response has been to strengthen its cooperative relationships in order to protect technological confidences and technical competencies; for instance it is seeking to increase its shares in Denso. Toyota has been strict about reducing the cost of components, something it has been criticised for in the past.

In the case of Nissan, its component suppliers largely carry out the design of the components. Hence if the suppliers were changed the designing would have to be taken over by Nissan, thus increasing the pressure on its design and development resources. Nissan is not going to discontinue its alliances but is trying to reduce the number of suppliers and improve the quality of its relations with these suppliers.

There have been similar developments in the US. General Motors used to produce 80 per cent of its components internally but now independent companies have been set up to produce them. GM allows these independents to sell components to other companies, but by the same token GM buys components from other suppliers if the quality and cost are better. Some components used to be procured by open bidding, but now they are bought from selected suppliers under long-term contracts (Dyer, 1996b), a practice apparently learnt from Japan. Recently computer supply chain management systems have been developed and suppliers that can operate this system are eligible to be selected as a member of the team. Hence the procurement system in the two countries seems to be converging.

The diffusion of the Internet is having an impact on the procurement system. Through the Internet, manufacturing companies can shop around for the cheapest component prices in the world. While this practice can be expected to spread, it will be limited to standard components.

Horizontal alliances

Before the Second World War, many alliances were formed to introduce new technologies to Japan from the US and Europe: NEC learnt much from ATT, Toshiba from GE and Mitsubishi Electric from Westinghouse. The war put an end to these relationships as the companies found themselves on opposite sides. After the war, many new licensing agreements were reached to introduce innovative products from overseas. Japanese firms were able to produce these products because they had the necessary technologies to do so. Sony bought the license for the transistor from ATT and invented small transistor radios, which began to be sold worldwide. Toray was able to buy the patent for nylon from DuPont by paying more for it than its net worth, and thus became a major exporter. Toray and Teijin bought the patent for acrylic fibre from ICI and became the largest acrylic manufacturer in the world. Matsushita learned much from Philips, as did Nissan from Austin (a British firm that despite having innovative products such as the Mini-Minor, a car that was very Japanese-like in that it involved the creative miniaturisation of key components, ceased to exist as an independent firm in the 1960s). Success in these alliances was due to two main factors: continuous learning and the ability to develop related technologies. The venture was not merely one of imitation.

Japanese firms attained worldwide competitive power in high technology products during the 1960s and started to invest in production facilities in foreign countries. Asian countries required such operations to take the form of joint ventures, but controlling power was vested in the Japanese firms because their parent companies held the key technologies and could develop them much faster than could their local partners.

CONTEMPORARY TRENDS

From competition to horizontal alliances

Horizontal alliances are increasing in number. For example Hitachi Kenki (construction machinery) and Furukawa Kikai (machine tools) are cooperating in the development of earth-moving equipment, Sony and Nihon Zeon (chemicals) are jointly developing a plastic hard disk with a much larger memory, and Hitachi and NEC are developing a large-memory semiconductor. Considerable concertation can occur. For instance six producers of stainless steel are cooperating to reduce overcapacity. Sony and Toho are constructing a large cinema complex that will show 13 films simultaneously. Soft Bank, Microsoft and Tokyo Electric Power are jointly setting up an Internet communication network. Honda is going to sell small diesel engines to GM, while GM will sell petrol engines to Honda. Two groups, comprising Matsushita and Toshiba on the one hand and Hitachi and Sharp on the other, are cooperating to construct plants for dismantling domestic appliances and recycling the components. The determining factors for these horizontal alliances are speed and economy of scale. During development, speed is most important, while in the production process, economy of scale is important.

SUCCESS AND FAILURE FACTORS IN ALLIANCES

Alliances are the subject of a considerable body of literature. For example Harrigan (1985), Contractor and Lorange (1988), Lorange and Ross (1992), Kanter (1994) and Lloyd (1997) provide comprehensive studies of strategic alliances, while the success and failure of alliances are studied by Rotter (1980), Kanter (1994), Mohair and Spekman (1994), Hitt and Ireland (1995), Hosmer (1995), Wathne *et al.* (1995), Mitchell and Singh (1996).

Going back to basics, Barnard stated that the principles of an organisation are three fold: sharing common goals, reducing communication barriers between members, and member commitment (Barnard, 1938). These three factors are also key to the success of alliances.

Other theorists would add other factors. For instance the theory of motivation, particularly 'expectancy–valence theory', places emphasis on the motivation to participate, on how partners maintain their alliance and when they should discontinue their association (Staw, 1976). Social exchange theory states that power arises where exchange is not equal. When the rewards given to one partner are larger than those received by the other then the former gains power and control. If one partner is less able or has fewer opportunities to abandon the relationship for another, then the other partner has the greater power (Blau, 1964; Chadwick-Jones, 1976). This theory investigates where power resides in reality, rather than merely reflecting the shareholding ratio. A partner with a minority of shares but with key technologies can control the other partners in an alliance. This is often the case with joint ventures in developing countries. Such alliances and ventures can often give rise to conflicts, many of which are productive and resolvable, but others are not (Deutch, 1969). Transaction cost theory seeks to explain the limitations of market trade. Williamson (1975) states that where there is opportunism, uncertainty and small numbers, markets will tend to be replaced by organisations. Game theory suggests that collaboration will occur when it is likely to result in larger pay-offs (Neuman and Morgenstern, 1953; Ruce and Raiffa, 1957).

Causes of failure

This section presents examples of the main causes of alliance failure:

- Financial loss.
- Opportunity loss.
- Goal differences.
- Differences in corporate culture.
- Lack of mutual trust.
- Lack of competitive power.

Financial loss

Financial loss is typically accepted as having occurred if, over a three-year period, a joint venture company, licensee company or firms in an alliance cannot make a profit. For instance, the supermarket Daiei tried to produce and sell colour televisions in alliance with a television manufacturing company, Crown, but they were not successful. As a sales outlet they lacked the competencies and credibility that the retailing of televisions requires. In such situations, both parties lose. Sometimes the losses are one-sided, but these are hardly any more sustainable. An example of this is a sales outlet being forced to hold too much stock in order to improve the appearance of the manufacturing company's annual financial return. Such relations do not last long.

Opportunity loss and failure to make an alliance

When the profits reaped by one side are much larger those obtained by the other side the alliance will not last long. An imbalance in pay-off usually causes trouble. Sometimes an opportunity loss may be sustained as a result of a trade limitation. Component suppliers in alliances are often not allowed to trade with assemblers in other groups, and licensees are sometimes not allowed to export products. In such cases dissatisfaction tends to arise. In extreme cases alliances are severed before the intended goal has been reached. Sony did not sell the patent for the Beta video player and consequently lost market share – this can also be seen as a case of alliance failure.

Goal differences

Matsushita formed a joint venture with a family firm in Bangkok to produce colour televisions and other domestic electrical appliances but the Thai partners seemed more interested in short-term profits and the promotion of family members. This policy was not congruent with the long-term goals of Matsushita and local managers did not appreciate the nepotism that existed. Hence Matsushita took steps to resolve the problem. Matsushita provided the key technology, so even though it held a minority share (49 per cent) it had the power to control the company. It separated the product division into independent companies and shielded them from the influence of its Thai partners. In addition the presidency and the managership of all six departments were awarded to Japanese nationals. In this way Matsushita regained control of the company and its performance is now considered a success. Conflict was avoided and the company continues to grow.

Hershey Foods and Fujiya (confectionery) formed a subsidiary in 1989. Hershey intended to enter the Japanese market by using Fujiya's sales channel and manpower. However its sales goal of 10 billion yen could not be attained. Sales had reached only two billion yen by 1992 because the new sales channel did not develop. Hershey did not trust Fujiya and decided to market independently, hence the joint venture contract was terminated.

Anheuser-Busch entered into an alliance with Suntory in 1981 with the intention of obtaining a market share in the alcholic beverages market of over 10 per cent. Yet after 10 years, it had only achieved 1.2 per cent. Suntory thought this was a good performance but Busch was dissatisfied and discontinued the alliance in 1995, starting a new alliance with Kirin Brewery.

Miller and Sapporo Beer entered into an alliance to sell Miller beer in Japan. A condition of the alliance was that Sapporo would not import any other brand than Miller, but Sapporo became dissatisfied with this and cancelled its agreement with Miller. Asahi Brewery seized this opportunity to enter into an alliance with Miller and a mutually beneficial agreement was reached – that is, Asahi would import Miller beer and in exchange Miller would allow Asahi to use its plants to produce its 'Super Dry' brand abroad.

Differences in corporate culture

In short-term alliances, such as those set up for research and development, differences in corporate culture do not matter very much, but it can be a problem in the case of long-term joint ventures.

Hitachi established an alliance with GEC in Wales to produce electrical products such as colour televisions. Production would take place in an old GEC television factory and Hitachi delegated the operation to GEC. However the factory was not kept clean and work practices were slack, to such an extent that workers could even smoke during assembly operations. Thus the company practiced incomplete quality control, product quality was low and the venture was unprofitable. Hitachi then took over GEC's share of the joint venture and introduced its own philosophy and culture, with enhanced emphasis on quality.

In an opposite example, Kirin entered a joint venture with Seagram to produce whisky. Kirin planned not only to produce whisky but also to learn the culture of Seagram. Kirin had the largest share of the Japanese market for beer and had been successful for many years, but its culture had become complacent and stagnant. Kirin wanted to learn from the more vitalised culture of Seagram through this joint venture.

Lack of mutual trust

Mutual trust means that each partner can rely on the other to act in its favour, even in difficult situations. Trust is an important cornerstone of successful alliances.

Teijin produced the polyester fibre Tetolon in the Philippines through a joint venture with a Philippines company. However the latter was suspicious of Teijin, saying that Teijin's newest technology had not being transplanted to the Philippines operation. For this reason the Philippines management introduced a technology from a German engineering company and changed the production process. When this did not work well and, despite remodelling, proved to be a failure, Teijin discontinued the alliance. In alliances, not all confidential details and elements of knowledge are necessarily transmitted. Partners need to understand this and be tolerant of it.

Alliances, even between very strong companies, fail if there is no trust in the relationship. Hitachi and Fujitsu were requested by MITI to cooperate in the development of semiconductors but the alliance failed to produce fruitful results because the two companies were rivals and did not enjoy mutual trust. The advertising firms Hakuhodo and McCann Erickson Worldwide launched a joint venture in Japan but the company they formed became a rival for Hakuhodo's own operations, so the latter dissolved the joint venture. Here the alliance 'gave birth to a wolf', to use the colloquial Japanese expression.

Lack of competitive power

Sometimes partners do not live up to what is expected of them in terms of contribution. Pfizer (drugs) and Taito (sugar) formed a joint drug-marketing venture called Taito-Pfizer in the expectation that Taito would contribute its sales power. However Taito contributed very little, thus Pfizer took over Taito's share.

Loss of competitive power is a frequent reason for ending a joint venture. Hitachi and Texas Instruments established a joint venture in Texas to produce semiconductors, but the plant was closed in 1998 without having made a profit. The same applies to NEC and Packard Bell, whose joint-venture PC plant closed in 1999 with a loss of 2000 job. Hewlett Packard also set up a joint venture with Yokokawa Electric to produce scientific instruments in Japan but sales declined, Yokokawa wanted to concentrate on semiconductor-related products and HP wanted to concentrate on PCs; thus the joint venture was discontinued in 1999.

Causes of success

The main factors in alliance success are:

- Common goals.
- Mutual trust.
- Complementary capabilities.
- Competitive and sought-after products.
- Mutual learning.

Common goals

Common goals are essential in alliances. Hitachi and NEC are competitors in the semiconductor business but they have been willing to cooperate fully in the development of large-memory semiconductors because this is a very expensive process. By sharing development costs and know-how they can gain a strong competitive edge over their competitors. Both parties expect to benefit in almost equal terms, hence their's is a win-win relationship.

Mutual trust

As discussed earlier, mutual trust means that each partner can rely on the other even in the most difficult circumstances. This trust is based on past experience and a number of assumptions: that the reward will be in proportion to the contribution; that each partner has a key resource to offer and will remain committed to the alliance even in adverse situations by maintaining good mutual communications (Hosmer, 1995).

If there is trust, companies will cooperate even when there is a partial conflict of interest. For example Fuji Film and Xerox formed a joint venture called Fuji-Xerox. Xerox has been learning from Fuji-Xerox about quality control and production processes, which were transmitted from Fuji Film, a case of benchmarking. Likewise Fuji is learning from Xerox through highly competent people it sent to Fuji-Xerox to cooperate with the development of new products. Toyota and its competent suppliers also have a strong relationship of trust. For instance when a component supplier has been faced with a decline in demand for a component as a result of shifting consumer preferences, Toyota has ordered other types of component to help ease the situation.

Setting up a cooperative venture involves a high degree of trust. For instance the alliance partners have to provide enough capital, competent personnel and the necessary knowledge to carry out the joint business. In the case of NUMMI (a Toyota joint venture with GM in the US), Toyota spent 40 billion yen on modernising the production facilities and trained 250 American foremen by sending eight groups (about 30 members in each group) to the Toyota plant in Toyota City for three weeks each. Toyota also sent 180 Japanese leaders to Fremont in California to help with the transplantation of the Toyota production system. It was this sort of investment in facilities and training that helped to create the trust between Toyota, GM and the trade union (UAW) that led to the success of NUMMI.

Complementary capabilities

Complementary capabilities and economies of scale can lead to worldwide competitive power. For example Sony provides the hardware for its PlayStation while its software suppliers provide outstanding software. In this way the PlayStation quickly became a highly successful product, whereas Matsushita had taken quite some time to organise its software suppliers when it entered the games machine business.

Seven Eleven Japan has about 7000 franchisees in Japan. The franchisees find stores in appropriate locations, appoint managers and provides funds, while Seven Eleven provides knowledge of store layout and the branding of the store, recommends merchandising, point-of-sale and information systems, and provides consulting services. The system proved so successful that it was exported to the US when Seven Eleven Japan acquired Southland, which operated 6000 American stores.

Many alliances are aimed at achieving economies of scale. The exchanges of OEM between the ballbearing manufacturers Nihon Seiko and NKK was intended to achieve an economy of scale by concentrating production on fewer items. The joint use of production plants by chemical companies is another example.

Competitive and sought-after

The products of alliances need to be competitive and sought-after if they are to be sold worldwide. NUMMI and Fuji-Xerox have been successful because their products are strong. Seven Eleven Japan has been successful because its merchandising is outstanding.

Mutual learning

Alliances can tend to create a sheltered, secure and stable environment because of the members' monopolistic position. The case described above of the telephone manufacturers in the NTT group is an example. Alliances should increase the opportunity for initial mutual learning as in the cases of NUMMI, Fuji-Xerox, Ford and Mazda. Not only should current knowledge be passed on but each partner should acquire and generate new knowledge as the environment and technology changes. Seven Eleven Japan updated its information system many times but Seven Eleven US did not and was losing money, thus Ito-Yokado (the parent company of Seven Eleven Japan) acquired Southland (the parent company of Seven Eleven US) and introduced the Japanese store management system to ensure survival.

The Toshiba Silicone Companies (740 employees in 1998) is a financially successful joint venture by Toshiba and GE. It has close contact with its customers and produces about 1000 types of silicone. Toshiba started to research silicone in 1941 as a material for insulating transformers and other electrical equipment. Silicone comes in both solid and liquid forms and has a variety of uses. The company places great emphasis on research, with 100 out of the 740 employees being engaged in research. GE also carries out research in the US. All employees participate in quality control activities. In the interest of mutual learning, many engineers from GE spend three months at Toshiba Silicone and Japanese engineers spend time at GE. The success factors at Toshiba Silicone are in line with other cases where mutual trust has been established: competitive products, an outstanding management system, investment for research and mutual learning, and a strong financial performance (more than 10 per cent growth in sales every year).

In collaborations and alliances there may well be factors that remain undisclosed. Not all technologies are disclosed to partners and some knowledge will remain confidential. Coca-Cola has many franchise bottlers whose job is to process and bottle the condensed Coca-Cola syrup, the recipe for which is never disclosed. The bottlers understand this and make a profit by working at the lower end of the value chain.

GE has many joint ventures in Japan, not only with Toshiba but also with Yokohama Electric, Funac, Sanyo, IHI and others, but it does not disclose its confidential designs. When one of the authors visited a joint venture by Asahi Glass in Indonesia that was producing plate glass using the float glass manufacturing process, the manager informed him that some of the confidential technologies had not been made available. Alliance partners should understand and accept this practice or, as illustrated by the case we related earlier of the Teijin polyester plant in the Philippines, the joint venture can fail.

SUMMARY

Successful strategic alliances combine the strength of two or more companies and create a core competence that cannot be attained by one company alone.

In Japan, vertical alliances or *keiretsu* are very popular and have supported the rapid growth of many high tech companies and enabled parallel enhancement of the technology used by component suppliers and assemblers. The car industry and the domestic electrical goods industry are the best examples of this. The group orientation of Japanese firms, as opposed to independence, has supported the growth of alliances. Alliances are also useful for international technology transfer.

Recently the number horizontal alliances has increased. These are set up to carry out expensive research, attain economies of scale or create de facto standards.

Key factors in alliance success are (1) mutual trust, (2) the existence of first-class complementary capabilities to create competitive core competencies and (3) continuous mutual learning to enhance the core competencies. Alliances enable firms to develop capabilities that cannot be developed by a single firm, to bring about economies of scale, to outclass competitors by establishing de facto standards, and to avoid the risks inherent in large individual investments.

Problems with alliances are that they create entry barriers, produce laziness on the part of protected partners or even 'give birth to a wolf' if one of the partners becomes too strong.

6 Multinational Management

CONCEPT AND TYPES OF MULTINATIONAL MANAGEMENT

Multinational management involves the transplantation of domestic management skills (usually often assumed to be superior) to foreign countries and building production bases there in order to increase market penetration, sales and profits. Superior skills in production management and the management of organisations are required to achieve competitive power in the host countries and the world market. Becoming a multinational is different from entering into exporting, licensing, plant engineering or the opening of sales branches of a company, in that they do not involve the construction of production facilities.

Depending on the aims of overseas subsidiaries, the latter can be classified into four types (Dunning, 1988). First, import-substitution subsidiaries (or market-oriented subsidiaries) produce the parent company's products to sell in the host country. The Asian subsidiaries of Japanese electronics products companies tend to belong to this type.

Second, production-centre subsidiaries (or production-efficiency-oriented subsidiaries) aim to export most of their products from the host country to other countries. In Asia, Toray's synthetic fibre company in Penan, Minebea's miniature bearing company in Shanghai and Canon's toner cartridge company in Darien are all examples of this strategy. These companies seek to achieve world competitive power by specialising in one or two products and achieving economies of scale, and are strongly controlled by their head office in Japan. Production in Europe and the US is also specialised because these markets are large and unprotected by import tariffs, so firms in these regions function as though they are production-centre subsidiaries.

Third, resource-oriented subsidiaries invest large amounts of capital to obtain natural resources for Japanese industry. Examples of such ventures are coal and copper mines and the extraction of iron ore, and ownership is often spread over two or three companies that share the costs and risks. Host countries such as Australia, while appreciative of the export earnings that this type of activity provides, would prefer companies to invest in processes that increase value added and boost employment, rather than simply leaving behind a large hole when the resources expire. However this does not sit well with Japanese investors: they want raw materials for their industries, not to build up the economies of potential competitors.

Increasingly, however, host-country concerns are shaping policies, as illustrated by the fourth type of subsidiary: economic cooperation subsidiaries. The steel production company established by Nippon Steel in Malaysia is an example of this. The company produces 200 000 tons of steel a year using local iron ore and charcoal derived from gum trees. When the plant became profitable the majority of shares were sold to the host-country government, delivering management to local managers. Nippon Steel developed a best practice example of economic cooperation and obtained the goodwill of Malaysia in the process. In the long term, type-three subsidiaries should adopt this practice if they wish to ensure their survival and continued success. However it will be difficult if the resources are extracted from remote locations, as then it will be much cheaper to export them than to develop and maintain expensive infrastructure in the host country, particularly when the host country is remote from the remainder of the Japanese value chain, as is the case with Australian ore and coal.

TWO TYPES OF STRATEGY

Multinational production can be classified as multidomestic or global (Table 6.1) (Porter, 1986; Flaherty, 1996; Clarke and Clegg, 1998). Multidomestic production is mostly carried out by import-substitution subsidiaries. The market is limited to the host country and the products are adapted to fit in with local needs. Global production involves the production of standard products for the global market by production-centre subsidiaries such as Minebea's Shanghai plant, which produces small ball bearings for the world market. In both cases the technology is transferred from the home country, but as production-centre subsidiaries tend to be large some research activity is carried out there. Some Japanese firms, for example, Toyota and Canon, have research laboratories in the US, but most Japanese companies conduct their research in Japan, where there is a large supply of university graduates in engineering and science.

A two-way transfer of personnel takes place between the host and home countries. In the case of multidomestic enterprises, managers tend to be from the host country because products and services have to

Table 6.1 Two models of multinational production

	Multidomestic	*Global*
Market	Host country	Global
Products	Fit local needs	Standard products
Competitors	Companies in host country	Companies through the world
Transfer of technology	From home country	From home country or two-way
R&D	Home country	Home country, but some local laboratories
Production plant	In host countries	Located where country-specific advantages exist
Personnel	Most managers from host country	Managers from home country, foreign managers at head office in home country
Advantages	Quick response to local needs, avoids high import tariffs	Utilises country-specific production advantages economy of scale.
Disadvantages	Little economy of scale	Poorer fit with the particular needs of different countries
Examples	Commerce, finance, cement, construction materials, import substitute production	Semi-conductors, ball bearings, PCs, tyres, watches

Notes:
Global strategy is quite different from a federation or subsidiary type.

Multi-domestic is similar to Clegg and Clarke's (1998) 'Multinational' and 'global' and 'transnational' are included in global in this chart.

Sources: Porter (1986); Yip (1989); Flaherty (1996); Clarke and Clegg (1998).

be adapted to local needs. For global products, however, a high degree of integration is necessary as technology is transferred from the home country, so the key posts are filled by home-country managers. When the scale of these subsidiaries is large, some foreign managers may be sent to work at the head office in the home country.

Among the merits of multidomestic production is rapid delivery, as in the case of YKK, a manufacturer of zip fasteners, which operates factories that meet the specific needs of the host countries and offers quick delivery. The merits of global production is that it is carried out in the best-situated places and it enables firms to utilise economies of scale. Global production is quite different from a 'federation' of worldwide subsidiaries, each of which engages in independent activities (see Flaherty, 1996). There are no such companies in Japan today: Yaohan, a department store operator, adopted this policy but went bankrupt because it failed to develop competitive core competencies.

SURVEY DATA ON PERFORMANCE

We carried out a survey on the effect of foreign direct investment (FDI) on performance. Companies were classified into three groups according to their percentage of foreign production over total consolidated sales: highly multinational companies (with over 20 per cent); moderately multinational (8–19 per cent), and slightly multinational (less than 7 per cent). As can be seen from Table 6.2, the highly multinational companies had considerably larger sales figures than the other two groups, indicating that FDI contributes to company growth. The difference in the ratio of foreign direct investment among the three groups was very large, even among manufacturing companies. The export ratio of the high FDI group was much larger than that of the other groups, indicating that high technology products from MNCs are highly competitive in the world market. Foreign production increases the export of components and the penetration and diffusion of a company brand: the fact of this high export ratio demonstrates that the FDI does not reduce employment in the home country.

The ratio of research and development expenditure over sales for the high MNCs was much larger than for the other two groups as technology-intensive products are most appropriate for foreign production. The ratio of sales promotion expenditure over sales for the high MNCs was much lower than for the other groups – the ratio of head

Table 6.2 Foreign production ratio, other strategies and performances of the 205 companies in the sample[1]

	Sales (parent company) (billion yen)	*Sales (consolidated) (billion yen)*	*Overseas production ratio*[2] %	*Export ratio*[3] %	*Research ÷ sales* %	*Sales promotion ÷ sales* %	*Staff in head office ÷ total personnel* %	*Staff in head office ÷ total personnel consolidated (%)*	*Growth rate of sales* %	*ROI*[4] %	*Standard deviation of ROI* %	*Equity ratio* %	*Overall performance*[5] %
High MNC (M ≧ 20%)	73.2	103.3	32.7	31.3	4.9	2.4	8	3	2.5* (3.2)[6]	5.8 (2.5)	2.8* (3.2)	41.6* (16.0)	12.2 (5.7)
Medium MNC (20 > *M* ≧ 8%)	44.7	63.9	12.0	19.1	4.3	2.8	9	8	1.7* (3.2)	5.7 (2.9)	2.6 (1.5)	36.2 (16.7)	11.1 (5.2)
Low MNC (8 > M)	42.8	54.0	2.6	10.0	3.1	4.7	12	7	2.1 (4.1)	6.1**[7] (4.2)	2.0* (1.2)	34.0* (16.5)	12.1 (7.4)
Average standard deviation	–	–	–	–	–	–	–	–	2.13 (3.58)	5.81 (3.32)	2.45 (2.09)	37.25 (16.61)	11.90 (6.29)

Notes:
1. For the composition of the sample see the appendix to chapter 1.
2. Overseas production ratio (M) is computed over the consolidated sales, 1993 data.
3. Export ratio is computed over the unconsolidated sales
4. ROI, its standard deviation and sales growth are computed from 1983 to 1993. ROI = profit before tax and interest divided by total assets, unconsolidated.
5. Overall performance = ROI – 1/2 standard deviation of ROI growth rate (≧0)+1/4 *equity ratio*.
6. The numbers in parenthesis are standard deviations.
7. Difference from average performance, ** < 5%, * < 10%.

Table 6.3 Foreign production by Matsushita Electric, 1996 (billion yen, estimated)

		Sales		
	Production	*Home country*	*Foreign country*	*Total*
Home country	5950	3600	2350	5950
Foreign country	1750	500	1250	1750
Total	7700	4100	3600	7700

Note:
This table does not include the amount of production of components manufactured in the home country and exported to foreign factories. Production in the home country would then have been larger than this table. In order to carry out this analysis, an input-output analysis would have to be used.

office personnel over total personnel was lower for the high MNCs, demonstrating scale effects. The high MNC's sales growth rate (unconsolidated) was higher than that of the other groups. Foreign production contributed most to the growth of sales.

The ROI of the high MNCs was rather lower than that of the low MNCs, while the standard deviation was larger. Companies have yet to harvest the fruits of foreign direct investment. (Similar findings can be found in Baden-Fuller and Stopford, 1988.) The equity ratio (unconsolidated) of the high MNCs was the highest among the three groups. Generally speaking the equity ratio is high when ROI is high (because retained profit is large) and low when the growth rate is high (because of a larger amount of debt finance), something that Table 6.2 does not explain.

Table 6.3 presents data from Matsushita Electric, which has 116 foreign production bases. Its ratio of foreign production over consolidated sales was 24 per cent in 1996. It has 150 000 employees in Japan (including domestic subsidiaries) and 150 000 in its foreign affiliates (including sales offices). The sales in foreign countries are supported by foreign production and account for 47 per cent of consolidated sales.

FEATURES OF JAPANESE MULTINATIONAL MANAGEMENT

Successful multinational management involves the transplantation of the core competencies and practices of the parent company to its foreign subsidiaries. (see also Ueki, 1982; Ishida, 1984; Ishikura, 1988; Yoshihara *et al.*, 1988; Tanno, 1993; Dunning, 1988).

Long-term orientation

Japanese firms do not seek profits immediately after an investment. The first three years are considered as a preparatory or experimental period, after which they expect to make a profit. This willingness to wait sometimes causes conflict with foreign partners in Asian countries.

Clear corporate philosophy

Many successful companies make their corporate creed clear to all employees of their foreign affiliates and key phrases are posted on the walls (see Chapter 3 for a full discussion and examples). Morning meetings are held in each section, at which the manager infuses the corporate philosophy, outlines the required performance in terms of annual policy, and addresses current issues. At Matsushita's affiliates the employees sing the company song and recite the corporate creed. After this an employee delivers a speech on his or her opinions or experiences. These practices help employees to understand the meaning of working for the company, as the company would have them see it, and increases their identification with the job and the company.

Toyota and other car manufacturing companies, Canon and other precision mechanical products companies, Toshiba, Hitachi and other electronics products companies have a high R&D ratio and high foreign production ratio. As already examined, 72 high FDI companies among the 205 manufacturing companies surveyed, have higher R&D ratios (see Table 6.2). This differs from other countries, where marketing intensive products companies, such as Danon, Heinz, Nestlé, McDonald's, Coca-Cola, tend to have higher foreign production ratios.

Transplanted production technology

In order to produce high-quality products, Japanese production equipment and modern automated machines are installed in overseas plants. For example, at NUMMI Toyota spent about $2 hundred million on robots and the modernisation of equipment, and Bridgestone systematically modernised its Tennessee plant after buying it from Firestone.

Emphasis on high quality

Not only the facilities but also the operating system affect the quality of products. Japanese firms usually insist on cleanliness in the interests of quality. At the Matsushita plant in Wales the slogan 'cleanliness creates high quality' is displayed in workplaces, while the 5S slogan (*Seiri*, neatness; *Seiton*, put in order; *Seiketsu*, cleanliness; *Seiso*, cleaning; and *Shitsuke*, discipline) can be seen in many Japanese plants. At Sanyo's plant in Dalien, China, 15 minutes are spent cleaning the workplace at the end of the day's operations.

In order to maintain the quality of products, key parts and key materials are imported from Japan. Steel plates are imported for various uses in machinery and copper pipes are imported for air conditioners. Moving components such as compressors for air conditioners, automatic transmissions for cars and cathode ray tubes for televisions are all imported. These imports serve to increase the exports of the parent company, as seen in Table 6.2. (However there is a growing tendency to produce key parts in the host country.)

Worker participation in quality control

The Japanese quality control system seeks to improve quality not by inspection but by worker participation in quality circle activities and suggestion systems. At Bridgestone's Tennessee plant training in quality control was provided by consultants and quality control activities were introduced. Worker participation gradually increased, and the best team was sent to Tokyo to attend the 'all Bridgestone QC circle meeting' (Ishikura, 1988). At Sony's Alsace plant (with about 2000 employees in 1991) there are 150 quality circles studying quality and cost. At the Denso plant near Melbourne (about 300 employees), all employees participate in quality circle activities. Suggestions for the improvement of working practices are encouraged and display panels encouraging people to make suggestions are widely displayed.

Although the number of quality circles and suggestions made outside Japan is much less than is the case in plants in Japan, Japanese subsidiaries in developing countries tend to have more quality circle activities than in Western countries.

Job flexibility and wages

Japanese subsidiaries try as much as possible to ensure job flexibility. For example at the Honda plant in Ohio there are only two job classes – leader and associate (although there are some specialist tasks, such as maintenance) – and wages are determined by these two classes. At NUMMI there were once 90 job titles but all titles have now been abandoned and employees are assigned to various jobs. In most subsidiaries wages are decided by the job. In Asian countries, where such wage systems have not been established, pay is sometimes aligned to status grade, as in Japan.

Equal treatment

In all Japanese subsidiaries, office workers and their heads of department share an open-plan office, thus improving communication between colleagues. There is no differentiation between white-collar and blue-collar workers – all wear the same uniform, including the director of the plant and the president of the subsidiary. The president frequently visits the production line and uses the same canteen as the other employees. At the Sony and Bridgestone plants in the US the presidents' visitors also take lunch in the canteen, sometimes together with the union leader.

Japanese subsidiaries try to protect jobs and do not readily lay off employees. For this reason there are many applicants for vacant positions, sometimes more than a hundred for one job. (In China the initial employment period is, for statutory reasons, about three years, but employment is usually extended.)

The equal treatment system does not apply to the remuneration of Japanese expatriates, who receive the same salary as they would in the home country, plus an overseas allowance. From the viewpoint of their lower-paid colleagues in the developing countries, this can be seen as unfair.

TRANSFERABILITY OF JAPANESE SYSTEMS AND PRACTICES

Not all elements of the Japanese system are transferable. In terms of acceptability to the people of the host country, the slow promotion of managers and ambiguous job definitions are hard to transplant, although flexible or multiskill jobs can be transplanted. There are a number of factors that affect the transferability of Japanese systems. These may be thought of in terms of five simple principles: (a) increase of labour productivity has to be accepted in any country that is to benefit from Japanese inward investment; (b) whether labour productivity touches the core value of the society; (c) whether there are conflicting values; (d) whether there are established systems; (e) whether there are spare resources.

The possibility that imported Japanese technology will increase labour productivity helps ease its acceptance. Most host country governments welcome such technology, as do employees, who can expect better wages through improved labour productivity.

In countries where individualism is strong, it is hard to transplant small group activities, the singing of company songs and a single, company-wide uniform. Where conflicting values exist between the host and home cultures, elements of the system that are congruent with a particular host value can sometimes be transferred, even where, overall, there is not a good fit. Usually, the more technically rational elements are easier to translate than are the more socially specific. These can present problems: for instance, Mitsubishi in the US was subject to a class action for its tolerance of sexually discriminatory practices. The Japanese style is to posit equality as a common value – but it is largely an equality of men. Using the same dining room for open-plan offices and the same uniforms for office and manual workers may be transferable.

It is difficult to transplant all elements of the Japanese wage system to the US as the two countries have quite distinct systems (see Chapter 10 for a full description of the Japanese system). However, it is much easier to apply in Indonesia, where there are several alternative systems that are premised on totally different principles and are not firmly entrenched. At issue is whether employees experience the difference as too great. In Britain, Japanese subsidiaries use the job classification system with a wide range of rates. While this is similar to the American system it is very different from the traditional British one, based on a flat rate for the job. The American wage system lies somewhere between the Japanese and British systems, so the change from a job rate to a job classification system is not perceived as very large.

It is difficult to establish sophisticated research laboratories or design departments in developing countries, and where the level of education is not high it is not as easy to introduce employee participation practices such as quality circles or suggestion systems.

Even within the same country there can be differences in the degree of transplantation (although in developing countries these differences tend to be small because advanced methods of management have to be applied at all subsidiaries).

There are two explanations for this. First, if an acquired company is relatively successful the management style will not be changed immediately and very few Japanese personnel will be sent to control the operation. However, if a newly acquired plant is not successful the management system will tend to be changed immediately, as in the cases of NUMMI and Bridgestone in Tennessee. Second, the degree of transplantation depends on the philosophy of the company in question. If the corporate culture of the parent company is well developed transplantation will tend to be extensive.

CONTEMPORARY TRENDS

Shift from import-substitution production to production for the global market

The a shift from multidomestic production to global production is somewhat similar to Vernon's (1966) life cycle model. Matsushita Electric had a number of subsidiaries in Asian countries, each producing a range of products – such as colour televisions, air conditioners, VCRs and batteries – for the host country market and importing the key parts from Japan. The import tariffs for these finished products was high, so even if the production costs were high the products could still compete with legitimate imports (but not with illegal, smuggled imports). The current trend is towards more specialised subsidiary production, made possible by lower import taxes and the establishment of free enterprise zones. Recently Matsushita established a production centre for cathode-ray tubes in Singapore, plus a marketing and training centre. Likewise electrical components, compressors and motors are produced at Matsushita subsidiaries in Malaysia. Toray has a large factory for the production of synthetic fibre products in Penang and Minebea, which makes miniature ball bearings, has large plants in Thailand and Shanghai, while Canon has a production centre for copier cartridges in Dalian. Each of these are specialist centres. China is now home to many production centres because of its high-quality labour force, lower wage levels and stable industrial relations. Similar changes can be seen at Sony in San Diego, where production has shifted from colour televisions for the US market to cathode ray tubes for the world market. The Sony plant in Barcelona has become a production centre for colour televisions for the EU, and Nissan in Mexico now specialises in the production of small cars for the NAFTA countries. Sony in San Diego (3700 employees) now has more than a hundred development engineers while Sony in Barcelona (1600 employees) has about 40 development engineers to design TVs that fit the different broadcasting systems in Europe. Research and development activities are increasing in subsidiaries.

Preventing hollowing-out effects

As foreign production increases, protecting employment in the home country becomes an important issue. Many companies have not reduced their workforce in the home country while increasing their production in foreign countries. In part, this has been made possible by an increase in domestic production of components for export. As we saw in Table 6.2, companies with a high ratio of foreign production also tend to have high export ratios (over 31 per cent in our sample). Hence the greater the foreign production, the greater the exportation of key components. Another factor preventing hollowing-out is the retention of mother factories in Japan. These function as sites for the development of new products and new mass production systems, experiments with new machines, the development of maintenance systems for production equipment, and the training of employees of foreign subsidiaries. For example Minebea (the miniature ball bearing manufacturer), uses its Karuizawa plant and development centre as mother plants and more than 6000 employees from Asian subsidiaries were trained there between 1972 and 1998.

CONTROL SYSTEMS

Head office control of multinational subsidiaries

There are three kinds of head office structure. One is a functional structure, where the production department controls the company's production plants around the world, the marketing department controls local sales offices, and so on. Specialised companies such as car, tyre and synthetic fibre manufacturers use this approach. The second is the product division type: here the product divisions control the

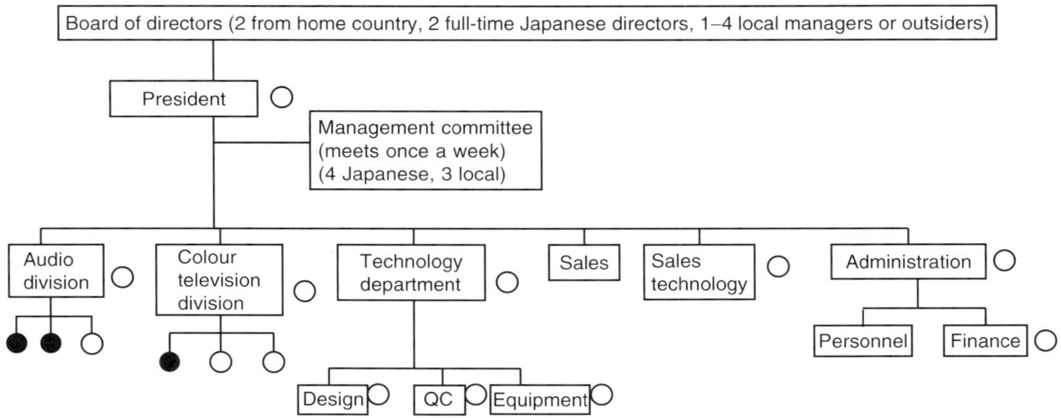

Notes:
1. ○ = headed by Japanese personnel. ● Non-Japanese
2. Total number of Japanese: 18 (including some not shown in the chart)

Figure 6.1 Organisation of subsidiaries, type 1: monopolisation of key posts by expatriates

company's foreign production and sales companies, as well as production and sales in the home country. Diversified companies such as Kobe Steel, Teijin and Komatsu use this approach. The third is the matrix type. Here the international department controls the area divisions, which in turn control production and sales, but the product divisions in the head office also control production and sales activities abroad. This is the most popular model in Japan (see Stopford and Wells, 1972).

Management and organisational structure of subsidiaries

The top management team at subsidiaries is usually a mixture of Japanese expatriates and local directors, irrespective of whether the subsidiary is a joint venture or a wholly owned corporation. The board of directors is composed of Japanese full-time directors, part-time directors from Tokyo and local part-time directors. The management committee (or a group of departmental heads), answerable to the board, is the actual decision-making body. It meets weekly or fortnightly and decision-making is by consensus. Group decision-making is a very common practice at Japanese subsidiaries and is a direct transplant of standard Japanese corporate decision-making.

In the case of joint ventures, controlling power may sometimes depend on the percentage of shares held, but even when Japanese ownership is less than 50 per cent the Japanese side has controlling power if it provides the key resources or technology. Thus controlling power depends on the amount of resources one of the companies provides, not on majority ownership. If there is a serious conflict of interests this ultimate power to control is important.

Four types of organisational control by Japanese expatriates

- Type 1: Monopolising key posts: this type is shown in Figure 6.1 and is found in production-centre subsidiaries, particularly during the start-up period. Japanese managers and engineers occupy all the key posts. There is strong control by the head office, and dissatisfaction among local managers often arises.
- Type 2: Assistant to department head: in this type the president and the heads of finance and plant are usually Japanese; the other top posts tend to be occupied by local people but Japanese managers work as their assistants. Matsushita Electric often uses this type of organisation. Local people can be promoted to higher positions. Where important decisions are made by Japanese employees, either

locally or with HQ, this causes dissatisfaction among local managers. Poor cooperation between local managers and Japanese assistants can also cause problems.

- Type 3: here Japanese employees are concentrated in engineering. This is often deployed by companies that produce technology-intensive products, including many of Sony's subsidiaries. There is less conflict between local managers and Japanese expatriates under this structure.
- Type 4: here only two or three employees are Japanese. From the Japanese point of view this structure tends to lead to failure, but in cases where an acquired company is already successful and remains economically viable it can work.

FACTORS IN SUCCESS AND FAILURE

Multinational management involves the application of company- and location-specific advantages to increase sales and profit by diversifying the location of production. If a subsidiary company is not making a profit but contributes to an increase in consolidated profit it can still be counted a success. This can happen when the transfer price from the home country to the host country is high. Commonly, however, success is measured by the profit level of the subsidiary.

Flaherty (1996) suggests that decision-making is the most critical factor in success or failure; however we argue that one must analyse the contents of strategy (Porter, 1986; Clarke and Clegg, 1998) (Table 6.4).

Selection of products

Japanese companies are strong in high technology products such as cars and consumer electronics, which have enjoyed considerable overseas success. On the other hand low technology goods such as laminated board and tinned pineapples have not been as successful. The Seiyu group (department store and supermarket) acquired a number of international hotels, as did many construction companies, but most of these did not make a profit. There was no synergy with the acquired businesses and the acquirers could not provide them with technological assistance.

Overseas manufacture is best suited to mass produced products. The products and production facilities are designed in the home country and the latter are exported to the host countries. This allows not only economies of scale but also easy control of the production process in the host countries. Bridgestone Tire has more than 10 factories in Japan and 20 overseas factories and almost the same production system is used at all of them.

On the other hand bespoke engineering such as shipbuilding is not appropriate for foreign operations because each vessel is produced to order and thus many engineers have to be located in the foreign country. Ishikawajima Harima Industries set up a shipbuilding subsidiary in Brazil in 1959 but it was unable properly to control its operation, never made a profit and divested all its shares in 1996. Likewise Japanese companies have not been successful in the foreign production of consumer non-durables – there are no Japanese equivalents to companies such as Unilever and Nestlé.

Selection of location

Host countries with location-specific advantages are those with a large marked for the product in question and an abundant supply of human and/or natural resources. For production-centre subsidiaries the size of market is less important than the human resources.

Failure occurs when a location is selected only because it has a supply of cheap labour. Many Japanese textile companies constructed plants elsewhere in Asia in the 1970s in order to take advantage of cheap labour, but most of them failed to make a profit because of their lack of product differentiation and overproduction. Few of them survived. Toray became successful by refining its technology and improving the quality of its nylon and polyester products, with the result that many competitors were driven from the market.

Transplanting the Japanese management system

A key factor in success is the ability to transplant company-specific advantages or management attributes such as long-term vision, strict adherence to quality, *kaizen* or improvement by participation, and respect for people. NUMMI is a typical success story.

The delegation of too much authority to local managers or other outsiders can cause failure. Toray's large synthetic fibre plant in Penang, Malaysia, is one such case. The construction and operation of the plant was delegated to a Hong Kong partner but the construction was poor, the floor was not flat, the machines did not operate properly and more than 10 per cent of products were defective. Later Toray took control of the subsidiary and reconstructed the plant,

Table 6.4 Sucess factors and failure factors

Principles of success		*Factors in Success*	*Factors in failure*
1. Transplantation of superior technology and management system		1. Selection of products: • high technology products • mass produced products • world-wide competitiveness	1. Products: • low-tech products • large bespoke products • no competitive strength
2. Economy of scale of the firm and of production		2. Transplantation of Japanese management systems	2. Too much delegation of authority or over-transplantation of home country system
3.1 Using location specific advantage 3.2 Merit of dispersion of location of production (saving of transportation cost)	→	3. Selection of location: • size of market • related industry • human and natural resources	3. Location: • Considering only low wage level
4. Contribution to host country		4. Contribution to host country	4. Disregarding the interest of host country and political move

changed the quality control procedure and transplanted high technology processes from the parent company. It became profitable in the 1990s.

The Nissan plant in Smyrne, Tennessee, is another example. There were fewer than 10 Japanese employees at this huge plant and problems arose in respect of vehicle defects. Nissan did not make a profit from this plant. Conversely the huge Honda plant in Ohio (with about 300 Japanese engineers) has been very successful. (For a different opinion on the delegation of authority, see Bartlett and Yoshihara, 1988.)

Benefits offered to the host country

In earlier times multinational companies such as the East India Company, were seen as invaders bent on exploiting the host country, but more recently almost all countries have welcomed them in the interests of economic development. Multinational companies contribute to the economic development of host countries by transferring technology to local component and material suppliers and by training local employees. They increase employment directly and indirectly, boost the export figures and pay taxes to the government. A number companies print brochures that explain how they contribute to the realisation of the economic development goals of the host country through these means, proclaim how the company devotes itself to the host country and list the production inputs that the company purchases in the host country.

Arabian Oil, a Japanese oil drilling company in Saudi Arabia, lost its major oil well when it was taken over by the government in 2000 because it was considered to be contributing nothing to the country. Likewise a fishing company in Indonesia had to close its operation because it was seen as overfishing rather than developing local assets.

Failure in decision-making

Full technical and sociopolitical information should be gathered before an investment is made. Some huge projects fail because they have been insufficiently analysed in advance. For instance the Mitsui Trading Company and four chemical companies in the Mitsui group set up the Iran–Japan Petrochemical Company in partnership with an Iranian public company in 1973. The intention was to extract petroleum and gas and to construct a petrochemical plant. About US$5 billion was spent over a period of 16 years but the project failed because there were insufficient deposits of oil and gas and construction was hindered by the Iranian Revolution and the subsequent Iraq–Iran War. The project was liquidated in 1989, although the construction was about 85 per cent completed (the site was bombed by Iraq a number of times).

PROBLEMS WITH OVERSEAS OPERATIONS

Interviews with employees, local managers, Japanese managers, and university professors in Indonesia, Malaysia, China and other host countries revealed the following recurrent problems.

Problems from the perspective of host countries

Japanese companies do not develop local industry

Japanese companies are often accused of picking the buds of infant industries in developing countries and of not contributing to the host country. Developing countries may want to develop their own motorcycle, car or synthetic fibre industries, or to modernise small businesses, but the presence of multinational organisations in the same line of business makes it impossible for them to do so. In response to this claim, Japan put restrictions on direct investment in the car, synthetic fibre and other industries until 1969 and successfully developed high technology products although licensing arrangements. Irrespective of the political economy in Japan, it can be argued that it has been successful because of a complex web of related technologies and the existence of many university graduates and highly skilled workers.

Developing countries can learn from multinationals. In the past, countries that accepted foreign direct investment, such as Malaysia and Thailand, enjoyed higher economic growth rates than India and China, which used not to allow entry to foreign multinationals. Successful multinationals can make it clear through their public relations activities how the affiliates increase employment, procure local components, give advice to suppliers and provide training. Yet in general Japanese managers in affiliates tend to be conservative about participating in social activities, such as speaking at meetings.

Japanese companies do not promote local employees to key positions

It is sometimes argued that Japanese affiliates do not promote local employees and that Japanese managers occupy all the key positions, while American corporations in Japan have promoted Japanese managers to all the top posts, including that of president – the presidents and almost all the managers of Fuji-Xerox and Japan IBM are Japanese. In other countries, local managers at American affiliates are more often promoted to higher ranks than is the case with Japanese companies.

As discussed earlier there are four types of organisational control, which reflect the degree of Japanese managerial domination, but in practice Japanese managers tend to control most affiliates. Over time local managers may be promoted to higher positions but some companies, such as National Thai and Toshiba in Tennessee, ensure Japanese domination through the monopolisation of key posts. Japanese managers often meet informally after work to make decisions and this 'shadow cabinet' is sometimes the most important decision-making body.

Discussion There are two reasons why Japanese managers dominate Japanese subsidiaries. First, local managers tend not to be proficient in the Japanese language and have difficulty communicating with the head office. Japanese managers of American subsidiaries in Japan, on the other hand, are able to communicate with the American head offices in English. Second, most Japanese manufacturing subsidiaries produce high technology products. Technological change is rapid and Japanese technical managers, rotated every three years, are important agents in the transfer of new technology to the affiliates. Many of the companies that delegated authority to local managers failed and had to change their system: for example Toray in Penang, GEC-Hitachi in Wales and Nissan in Tennessee. We should not forget that the survival and prosperity of subsidiaries due to the efforts of Japanese managers benefit employees and host country alike.

Japanese companies do not establish research laboratories in host countries

It is often argued that Japanese firms only import and assemble parts and do not engage in technological research in the host countries, an accusation also levelled against many other multinationals (Clarke and Clegg, 1998). However it must be recognised that research activities need high-calibre researchers and considerable resources. Even in Japan the major research laboratories are under head office control. Yet as foreign activities increase in scale and the degree of local procurement rises, development work is carried out to adapt products to local specifications.

Japanese managers are isolationist

Another common complaint is that Japanese managers mix socially only with other Japanese and never with local employees. In Japan, managers meet colleagues and subordinates after work for a meal or a drink, but in foreign countries they tend not to do this. In developing countries they often live in large houses in secure districts that are remote from the homes of their host-nation colleagues. Furthermore Japanese managers, it is said, look to Tokyo all the time and long to go back to the home country, while insisting on lifetime loyalty from locals.

Discussion: There are good reasons for what appears to be Japanese isolationism. In developing countries Japanese managers usually live in guarded areas for security reasons. Since Japanese people tend not to like living outside Japan, good housing is provided in compensation. They are also hesitant about going to local restaurants because of cultural differences in cuisine and food safety, and they do not enjoy risking their life in countries and cultures they do not understand. Furthermore in developed countries employees commute by car, so after-work drinks with local colleagues are not usually possible.

The rotation of Japanese managers is necessary to renew knowledge and for promotion, so Japanese managers do not stay long at foreign subsidiaries. Thus their situation differs from that of local employees. It would be a great mistake for top managers in Japan to advise staff going overseas to 'bury your bones in the host country'. On the other hand, while it seems desirable for local employees in the host country to be rotated among subsidiary companies in other countries in order to increase their promotion opportunities, this is not as easy as it appears for reasons of language and nationalism. Local managers in Asian countries are often not proficient in English (even in the Philippines English is not used for conferences), and some nationals tend not to welcome people from neighbouring countries. Moreover some, especially the older generation, still bear the scars of the years of Japanese occupation during the Second World War.

Problems from the perspective of Japanese managers

Loyalty

Local employees tend to be less imbued with company loyalty and will leave the company if they can get better pay elsewhere. Japanese companies provide considerable training, both on and off the job, and after gaining better skills employees are sometimes lured to another company by higher pay. This tendency is accelerated when the subsidiary applies a policy of slow promotion, following the Japanese style. To prevent this problem subsidiaries must introduce a policy of speedy promotion for competent employees. Some think that the transmission of skills to the host country is best achieved by ensuring that local employees do leave the company to work elsewhere, thus distributing their acquired expertise domestically.

Inflexibility

Local employees, it is said, only follow orders or restrict themselves to the specified job, they are not flexible when required and are more inclined to specialisation. Whether or not this is the case, to Japanese managers this often looks like individualistic behaviour and indicates less commitment to the company than is the Japanese norm. In order to prevent this problem, Japanese companies try to encourage quality circle activities and suggestion schemes. Also, in foreign subsidiaries the simplification of job grades is useful. At Nippondenso, the car air conditioning manufacturer near Melbourne, the 18 job grades were reduced to six and the number of job titles was reduced to three: production, store and maintenance associate.

Competition

Often in overseas postings, communication between Japanese and locals is poor. In addition local employees tend to be competitive rather than cooperative, and those higher up the status or job ladder do not pass on their knowledge to their subordinates as possession of this knowledge might lead to their subordinates being promoted faster than they are. From a management viewpoint, in order to avoid this problem and improve knowledge transfer, quality-circle activities might be useful.

Government control

In developing countries government control tends to be strong, and there are often many uncooperative bureaucratic departments that control some or other aspect of the firm's dealings. For example obtaining permission to import components can take a long time and components are often held up at the docks for days or even months. (In one country that we are familiar with, all documentary material had to be passed by the censor, whose staff were somewhat slow and books, manuals or instructional tapes could take months, or years to be passed.) In order to avoid this problem, many countries have set up free trade zones or industrial districts.

Expatriate lifestyle

The children of the Japanese managerial classes are expected to go to good high schools and good universities in Japan, hence the wives and children of managers tend to stay in Japan when managers are posted to a foreign country, where they live a single life. The rotation of expatriates is partly conducted for this reason (and partly for knowledge renewal), a tour of duty being five years for administrative staff and three years for engineers. As mentioned above, in developing countries expatriates live in large houses in special areas and this often causes envy among local employees.

Problems from the perspective of potential foreign investors in Japan

Foreign investment in Japan is much less than foreign investment by Japan. In 1994, outward direct investment by Japan in foreign countries was about 8 per cent of GNP but inward direct investment was only 0.7 per cent of GNP. By comparison, US and UK outward investment was 6 per cent and 10 per cent and inward investment was 5 per cent and 8 per cent respectively (Flaherty 1996; Japan Trade White Paper, 1997; Clarke and Clegg, 1998). (Note that direct investment includes manufacturing, commerce, finance and real estate but excludes investment in securities.) The reasons for this imbalance are as follows. First, the high cost of land, labour and energy in Japan deters inward investment. Second, the establishment of new sales channels is difficult. Third, competition is severe as Japanese products are highly competitive at home as well as abroad. Finally, and perhaps most importantly, strict government regulations have deterred entry by foreign companies.

While it might be argued that the Japanese state is seeking to maintain a fortress Japan unpierced by foreign competition, some companies, such as IBM, 3M, Hewlett Packard, Coca-Cola and a number of oil companies, have successfully invested in Japan. Japan has much to offer these companies, such as high-quality labour, a large supply of engineering graduates, a banking system awash with capital and a large domestic demand.

SUMMARY

Multinational management involves the construction of production bases in foreign countries and the transplantion of superior technology and management systems. There are four types of subsidiary: import substitution subsidiaries, production centre subsidiaries, resource-oriented subsidiaries and economic cooperation subsidiaries. The import substitution type tends to adopt a multidomestic strategy and the production centre and resource acquisition types tend to adopt a global strategy.

Our survey revealed that companies with a high rate of FDI also have a high R&D ratio, a high export ratio and a high growth rate, in which the return on investment is almost the same as for other groups. However the high FDI group's overseas production ratio was about 30 per cent, so if they had not engaged in foreign production their performance would have been very poor.

Japanese multinational production leads to technology-intensive products being produced and Japanese style management being transplanted. A long-term orientation, clear vision and emphasis on quality are important features of the management system. (That the price of Japanese used cars abroad is higher than that of other used cars is evidence of the success of this strategy: Japanese products have a reputation for quality.) Other features are participation in quality circles and suggestion schemes, and employment security and equal treatment for all employees, from the top to the bottom. Because of the latter there is a flood of applications by potential new recruits when a job falls vacant. The extent of transplantation depends on the social culture of the host country.

There seems to be a shift away from import substitution production towards production for the global market as the barriers to trade between countries are lowering. At the same time every effort is being made by multinationals to protect employment in the home country, for example by changing their original factory to a 'mother plant' for the development of new products and new production methods, and for training employees from the host countries.

For many firms, head office control of subsidiaries is organised as a matrix. Other subsidiaries are organised in terms of product divisions or as a functional organisation. In all cases the key posts tend to be occupied by Japanese managers. It is believed that this is necessary to produce high technology products. The delegation of authority to host country managers has tended to fail in the Japanese case.

Success factors in multinational management are the selection of suitable technology-intensive products and mass-produced products, and the transplantion of strict quality control systems based on staff participation. Failure occurs when low-technology products are made in countries with cheap labour and when no contribution is made to the host country or political problems are disregarded.

Problems from the host country's perspective include the fact that host-country nationals are not promoted to key posts, and that multinationals only assemble imported components in the host country and do not carry out research and development. From the perspective of Japanese subsidiary managers, local employees are insufficiently loyal and tend to leave the company when they are offered higher pay elsewhere, and this often happens after an investment has been made in their training. Furthermore they only do the job they are ordered to and are unable to be flexible when required.

The flow of FDI into Japan is much smaller than the outflow. The reasons for this include the high price of land, high labour costs and, in particular, restrictive government regulations.

7 Competition Strategy

CONCEPT AND MODELS

When many similar products are sold in the same market there is competition. Competition strategy seeks to increase or maintain the market share of established or new products. Competition takes place on four fronts: product quality and performance, cost and price, sales promotion and service, and sales channels. Combining these leads to four competition strategies: product differentiation, market segmentation, price policy and cost leadership, and construction of entry and mobility barriers (competitive strength has five levels of hierarchy, as shown in Figure 7.1, a model similar to the upper part of the value chain model developed by Porter, 1985).

The past performance and goals of a company generate the other levels. Past performance, particularly accumulated profit, is an important resource for competition. First, long-term, aggressive or high-level goals affect competition strategy. Japanese companies place

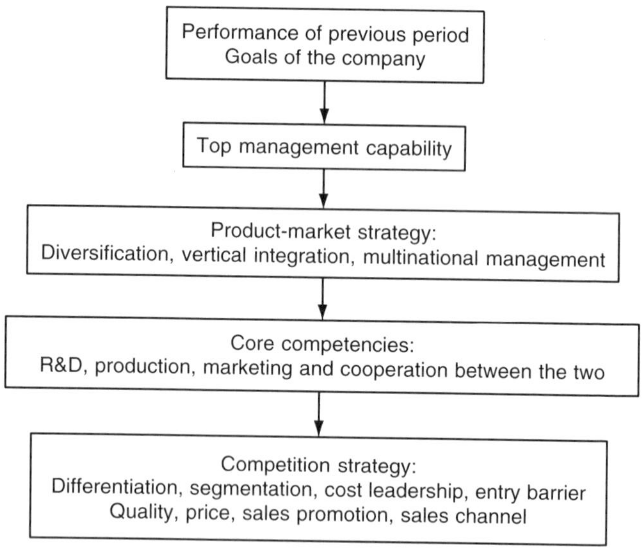

Figure 7.1 Hierarchy of competitive strength

considerable importance on the long-term growth of sales and thus tend to neglect short-term profits. At the second level, the top management makes decisions on strategy and structure. At the third level, product-market strategy is the basis of competitive strength. Synergistic relations with other products and time series or dynamic synergy) affect this competitive strength, as when strength in the production of electronic consumer products leads to competencies in semiconductor technology. Multinational production drives down manufacturing costs to achieve cost leadership in high technology products. For example Aiwa produces more than 80 per cent of its electronic appliances in Asian countries and this strategy is the source of its cost leadership.

Profitability and market share

Competitive strength can be measured by market share. We examined the impact of market share on the profitability of 205 companies, using multiple regression analysis:

$$\begin{array}{llllll} \text{ROI} = 6.69 & -3.42\,\text{PMT} & +0.69\text{PMT}^2 & +0.41\text{BGT} & +3.44\text{SHA} \\ \quad (1.48) & (1.46) & (0.32) & (0.07) & (0.61) \end{array}$$

$$R^2 = 0.29,\ F = 20.21$$

(ROI represents profit before tax and interest divided by total assets, 10-year average for the period 1983–1993. PMT represents level of diversification, S = 1, D = 1.3, RMT = 2, RT and RM = 3, U = 4. For definitions, see Chapter 4. BGT represents the sales growth rate [unconsolidated] between 1983 and 1993. SHA represents market share. Largest company = 1. The following formula was used for the proxy variable: consolidated sales of one company divided by the sales of the largest company = share. The numbers in parenthesis are standard deviations.)

According to these results, market share has a considerable effect on profitability. Of the other factors, diversification has a U-shaped effect and growth rate has a positive effect.

FEATURES OF JAPANESE COMPETITION STRATEGY

A comparison of the Japanese and US competitive strategies is presented in Table 7.1. Japanese firms are long-term oriented because

Table 7.1 Comparison of features of the Japanese and US competition strategies

	Japan	*Traditional US*
Goal	Long-term growth	Short-term profits
Investment	Emphasis on R&D and employee training	Less capital investment
Cooperation between suppliers, makers and sales channels	Good cooperation between suppliers, manufacturers and vendors (simultaneous engineering, JIT system)	Open bidding, independent sales channels (long development time, large inventory)
Interface between departments	Good interface (simultaneous engineering)	Poor interface between development, production and sales departments
Focus on means of competition	Quality	Cost (there is an economic quality level)
Customer relations	Customer oriented	Mass production, mass sales
Attitudes	Me-too, strong competition	Niche seeking

Sources: Dertouzos *et al*. (1989); Clarke and Clegg (1998).

shareholders have little power and core internal-labour-market employees enjoy lifetime employment. Large investments in research, development and the modernisation of equipment support the firms' competitive strength in high technology products. Sony and Canon spend about 10 per cent of their sales receipts on research and development.

Extensive and effective cooperation exists between assemblers, component suppliers and sales channels, irrespective of whether or not the assemblers control the latter two. By contrast, in the US car industry most components are produced internally and the rest are procured by open bidding. Such relations are less conducive to cooperation between the parties (see Dyer, 1996b, but note that this system is changing and becoming more like the Japanese one). The Japanese car industry, because of quasi-vertical integration and the use of simultaneous engineering, requires component manufacturers to participate in the design process from an early stage. For example moulds, parts for press machines and the transmission are designed at the same time as the body and engine. The component suppliers not only join in concurrent engineering but also make suggestions on new technology and try to cut costs by value engineering. All this serves to reduce the cost of production and the development time.

Changing the production line when new models are introduced has to be completed quickly. To this end, cooperation between the design, production and marketing departments is essential, so the development team comprises members from all three. The design engineers visit the plant to determine how to reduce the labour involved in production. Such cooperation reduces the lead time between designing and starting the production line. This practice is in sharp contrast to the one where development is carried out like a relay race, with the baton being passed from the development to the production to the marketing department. The barriers between such departments are relatively impervious (Dertouzos et al., 1989).

Product quality is the most important element of the Japanese competition strategy. For instance when Komatsu started to produce earth-moving equipment to compete with Caterpillar, the most important goal of the development team was to improve quality at any cost. Bridgestone had the same goal when it acquired the Firestone plant in Tennessee, modernised the equipment and retrained the employees (Ishikura, 1988). Because of the high quality they achieved, Komatsu and Bridgestone gained a large share of the world market. Likewise the high quality of Japanese cars resulted in their holding a 30 per cent market share in the US, and Japan now dominates the world market for small electrical components, such as condensers, resistors, and magnetic reading heads. High quality is attained not only by the use of modern equipment and tight quality control but also through the participation of employees, thus quality is produced through the production process. Quality-circle activities and suggestion systems contribute to this end. They also enter into design: Japanese products are highly consumer-oriented in design and have many customer-friendly features.

When a growth area emerges many companies wish to enter it, resulting in severe competition. In Japan there are eight car manufacturers, six companies in the semiconductor business and five domestic appliance manufacturers. Hence the Japanese competition strategy does not include niche seeking.

CONTEMPORARY TRENDS

Two worldwide factors are driving contemporary trends in Japan: the emergence of mega-competition following the lowering of entry barriers and changes to the rules of the game; and technological changes particularly in the case of information technology (IT). (See D'Aveni, 1995a on hyper-competition, and Clarke and Clegg, 1988, for an overview of the new competitive approaches.)

Agile management by supply chain management

One new trend that is gaining ground in Japanese manufacturing is truncated supply chain management. For instance in the clothing business the World Company owns many shops as well as operating boutiques in large department stores. Its strategy is to analyse the previous week's sales every Monday, and on the basis of this to forecast sales for the next two or three weeks. On Tuesday orders are placed with selected clothing factories, perhaps for new designs, and on Friday the following week the garments appear in the shops. This system differs from that of other clothing companies in that the latter tend to forecast fashion trends six months ahead and produce the goods for stock before selling them through department stores or other retailers. These retail outlets do not take a huge risk because they can always return unsold garments to the manufacturers, often resulting in a large stockpile of unsold clothes that are eventually sold at vastly reduced prices.

Producing to order is also taking place with high technology products. For example at Fujitsu's Ishikawa plant, standard PCs are partly assembled and sent to the Yamato Plant, where orders are received every hour from the head office. Here the workers assemble the remaining parts in line with the 'ticket' (specification) and add the required memory capacity, software and other specifications. One worker does the complete assembly for one product, which is then sent to the inspection line. The plant holds no stocks of finished products, but does hold a small supply of components, which are sent many times a day from the suppliers.

Generally speaking this made-to-order production utilises a 'cell production system' instead of a conveyer-belt system. In the cell production system either one person completely assembles the product or several people do a number of jobs each. Even when a conveyer belt is used, as in the case of cars, the length of the line is shorter: there are many short lines for a variety of cars, each with different specifications.

IT and production technology

IT is not only used for communication and documentation but is also applied to production technology. First, some successful Japanese firms have improved their capacity to produce components for IT hardware by applying IT to production, and now dominate the world market in LCDs, DVDs, RAMs and a number of components of mobile phones, such as small registers, condensers and transformers. They also dominate the world market in production machinery for IT-related equipment and components, in areas such as semiconductors, robots and moulds for components. For example the metal mould manufacturer Incs Inc (Japanese companies make more than 60 per cent of the world's supply of moulds and this company is one of the leading firms) uses three-dimensional CAD to envision a new product, such as the casing for mobile phones. Using numerically controlled (NC) machine tools, the new mould is produced in about six days, compared with about 45 days using the older method. Since mobile phone models change very quickly, having a short development time for moulds confers a strong competitive advantage. (This information was drawn from a presentation by S. Yamada at the Japanese Academy of Organizational Science's conference in May 2000.)

At Funuc Co. the production process of NC tooling machines is so automated that robots are produced by robot machines. Likewise at

NTN and Minebea, which produce high-quality ball bearings, the pressing, cutting, heat treatment and grinding processes are fully automated, although the assembly line is still staffed by operators. The production of hot coil steel is also highly automated. In this process the speed of the rollers that press the steel plate changes as the plate becomes thinner, controlled by computers that include a contingency program to cope with irregular problems that might occur. The reason why many steel manufacturers have entered into the production of semiconductor and computer systems is that they have considerable knowledge about the operation of computers, and Nippon Steel's software business is one of its main sources of profit.

Cost leadership through mergers and horizontal alliances

In a growth economy, many firms expand their capacity in order to meet demand and attain economies of scale. In a slow-growth or stagnant economy, economies of scale can only be attained by integrating separate production units.

Mergers

Mergers have taken place in unprecedented numbers in Japan in recent years and are likely to continue. For instance Oji Paper was born in 1996 through the merger of Honshu, Kanzaki and the original Oji, and a new paper company is due to emerge in 2001 as a result of the merger of Daishowa, Nippon and Jujo-Sanyo. It is predicted that the 20 car manufacturers in the world will eventually be merged into six groups, with only Toyota and Honda staying independent. Such mergers are horizontal, not attempts at diversification. After a merger the senior executive is selected alternately from the original pre-merger companies, in what is referred to in Japan as a 'crossing promotion policy'.

Horizontal alliances

In horizontal alliances the production units of competing companies are integrated to enable economies of scale. Hitachi and Mitsubishi Heavy Industries, which are competitors in steel production equipment, such as hot strip mills, combined their development and production activities by establishing a joint-venture subsidiary. Hitachi has also joined forces with another competitor, NEC, in order better to compete with European forces in Asian markets.

The above examples are of alliances between strong and highly competitive companies.

Alliances between weak companies

Konica and Minolta were both relatively weak in the production of copiers – to rectify this they entered into an alliance to develop and produce components and toners. When copiers changed from analog to digital these companies were at the forefront of the new technology.

Divestment by transferring the production of weak products is also a means of achieving scale economies. Toshiba theoretically divested itself of air conditioners, automatic teller machines, large capacity motors and semiconductors by setting up a joint production venture with Motorola. Such divestments are carried out by transferring production to other companies rather than by closing the plant and firing employees. Japanese companies try to save jobs even when contracting out production.

Impact of the Internet on vertical integration

The procurement of components and raw materials through the Internet will change the relationship between component suppliers and assemblers, and a price war can be expected. Toyota and GM are cooperating to open a virtual market for components from global suppliers. However key components and material such as steel plate will still be procured by long-term contract because their quality is so important. Even nuts and bolts are regarded as key parts because defects in these can cause serious accidents.

The purchase of cars, domestic appliances and books via the Internet is already well established in Japan. In 1999 Nissan sold 7700 cars through the Internet via dealers with which it was in alliance. Some consumer electronics, such as the PlayStation, are sold through the Internet by mail order companies. However companies with strong sales alliances cannot take this course because it erodes the position of their alliance partners. In this sense Matsushita (with more than 5000 allied sales outlets) has less freedom to sell on the Internet than Sony (with fewer than 1000 outlets, which also sell other companies' products). Internet sales may have some unforeseen effects on established competitive strategies.

The Internet is also a useful means of communication and advertising. Nissan has a virtual club where the company communicates with club members, and by this means the company can collect information on customer needs. Nissan's virtual showroom enables potential buyers to look at its cars and study their specifications.

Differentiation and de facto standards

Since hardware has to be supported by software, establishing the standard format has becomes a matter of fierce competition. A pioneering company can establish a standard with cooperation from other companies. If there is a hardware competitor with a different format, the standard adopted will depend on how many companies join each side. While Sony and Toshiba unified their DVD-ROM formats they did not unify their DVD RAM and are competing to attract adherents to their respective standards.

Japanese firms have developed many de facto global standards, including the VHS for VCRs and the format of CD players, game machines and digital cameras. The format of cellular phones has yet to be standardised.

PROBLEMS WITH COMPETITION

Being imitative and too competitive

Japanese firms are often accused of being imitative rather than creative. In addition many companies enter the same field simultaneously, causing severe competition. For example, in the semiconductor field NEC, Toshiba, Hitachi, Fujitsu and Mitsubishi Electric all produce DRAMs, the price of which fluctuates strongly in line with demand and supply. They all export the same item to the same country, destroying the market in that country with a flood of low-priced goods.

Discussion: Japanese firms spend a large amount of money on research. For instance the Japanese patents registered in the US in 1995 amounted to 21.5 per cent of the total compared with 6.4 per cent for Germany, 2.4 per cent for UK and 2.8 per cent for France (the US held 55 per cent: Japanese Patent Office, annual report, 1996). In the technology trade, in 1996 the receipts from Japanese patents were larger than payments made for foreign patents (Management and Coordination Agency Statistics, 1997). Japanese firms have pioneered many products, such as automatic cameras, DVDs and CD players, as well as processes such as the JIT system, benchmarking, quality circles and fish bone analysis.

Vertical integration acts as an entry barrier to foreign companies

Japanese distribution channels are very complicated and multilayered, and there has been much criticism of their efficiency.

Discussion: Many companies, like Toyota, have opened their sole distributors to foreign companies. In addition, small goods like electric shavers, PCs, cosmetics, film, and tennis rackets have not been considered appropriate commodities for integrated sales distribution in Japan. Manufacturer sponsored distributors are organised by the similarity of commodity not by needs of customers. However, tied distributors, such as those that work in alliance with Matsushita, are now losing market share and larger stores are increasing sales. In the vanguard of these new trends in the US toy retailer, TOYS 'Я' US, which has been successful in expanding in Japan through selling a group of products, children's goods, that meet the same needs.

Trade restraints

Companies are protected against imports and foreign direct investment as government licenses function as entry barriers to foreign companies. For example, at one time licenses for the computers used for the government offices and state universities were only granted to Japanese companies. Many such restrictions have been abolished since 1967 but they remain strong in the case of agricultural and dairy products, for example rice (as discussed earlier, Japanese consumers are very wary of non-Japanese rice, so in this case it is not just government policy that restrains trade). Cross-shareholding continues among *zaibatsu* group companies, which not only protects companies from hostile takeover but also impedes free competition.

Discussion: One should point out, however, that many of the government regulations that deter foreign trade are being abolished or eased. The number of mandatory licenses issued by the various ministries declined considerably in the 1990s, especially after 1996, and there is no longer a restriction on foreign direct investment in Japan. Foreign companies have acquired many financial institutions indebted as a result of past lending practices. We can differentiate two groups of industry. One group is regulated by the government and sometimes seen as a protected industry. Financial institutions are protected against bankruptcy. Thus, they over-invested. Another group is only minimally regulated and not protected by government. Manufacturing industry belongs to this group, where, if unsuccessful, firms go bankrupt. Thus, manufacturing industry has developed considerable competitive power in the world whereas Japanese financial institutions, despite being very large in terms of asset-values, have never been as competitive. However, non-government sources of protection are evident. Cross-share holding is decreasing in order to improve the profitability of investments but still protects companies against hostile takeovers. At Hitachi, foreigners held about 27 per cent of shares while 26 per cent of the shares were held by large, friendly, shareholders in 1998.

SUMMARY

There are four elements of competition – quality, price, sales promotion and sales channels – and four competion strategies: product differentiation, market segmentation, cost leadership and the construction of entry barriers. The strength of these strategies depends on the company's goals, the capability of top management, the product-market strategy and the organisational structure, as shown in Figure 7.1.

The competitive strengths of Japanese manufacturing companies have been well analysed by Dertouzos *et al.* (1989) and include long-term vision, emphasis on research, cooperation between the links in the value chain, cooperation between departments, emphasis on quality and a customer orientation. These are the result of companies' long time horizon and group orientation.

New practices have developed in response to mega-competition and rapid changes in technology. These include truncated supply chain management; the application of IT to numerous production processes, and to procurement and sales; horizontal alliances and mergers; and a race to set de facto standards.

Common criticisms or misunderstandings by foreign companies and researchers are that Japanese firms are too imitative and competitively aggressive; that vertical alliances and complicated sales channels act as entry barriers; and that MITI and other government departments erect entry barriers through licensing.

8 Planning and Decision-Making

Decision-making involves choosing between alternative courses of action, while planning involves goal setting, information collection, idea generation, evaluation and preparation for implementation, as shown in Table 8.1. The process of planning is usually synonymous with that of decision-making.

Decision-making relates to many areas but most typically concerns product-market strategy, structural design (for example capital investment and change of organisational structure) and operational planning (for example production and sales planning). Depending on the extent of change, planning may be classified as directed at innovation, improvement or maintenance. In this chapter we shall focus mainly on strategic decision-making by Japanese firms. (On the above definitions see Ansoff, 1965; Hofer and Schendel, 1989; Kono and Cudd 1998.)

Strategic decisions can be classified as (1) innovative and analytical, (2) innovative and intuitive, (3) bureaucratic or (3) conservative and intuitive (see Appendix 1). The decision-making style of successful Japanese firms is close to the innovative and analytical type. We shall examine this model in detail.

FEATURES OF JAPANESE DECISION-MAKING

The main features of Japanese strategic decision-making are shown in Table 8.2. As these features have already been discussed in previous chapters we shall consider them only briefly here (see also Linouwes, 1993; Ouchi, 1981; Kono, 1984a; Dertouzos *et al.*, 1989; Kono and Clegg, 1998; also relevant are the concepts of strategic intent, proposed by Hamel and Prahalad, 1989) and core competence, proposed by Prahalad and Hamel, 1990).

Goals

Japanese firms emphasise long-term growth rather than short-term profits because consideration of the interests of shareholders is minimal compared with that of banks and employees: long-term growth increases

Table 8.1 Areas of planning

	Innovation	*Improvement*	*Maintenance*
Basic goals and policies	Major change of business creed	Minor change of business creed	
Product-market strategy	New product development, vertical integration, new foreign direct investment	Competition strategy	
Structural planning	Capital investment, acquisition of an other company, strategic alliances	Recruitment of personnel, Improvement of facilities, improvement of rules and standards	Wage and salary system, Maintenance of facilities, rules and standards of production and sales
Production and sales activities			Procurement plan, production plan, sales plan, profit plan

Note: The subject of strategic decisions is mostly in the area of 'innovation'.

the promotion opportunities for employees and investment opportunities for banks. The banks exercise what Murphy refers to as 'credit rights':

> In Japan, an elaborate, although informal – that is not codified – system of credit rights is administered by powerful officials and unofficial bureaucracies. Credit is thus allocated through criteria that are fundamentally bureaucratic in nature rather than market driven – an institution or person that meets certain criteria receives a right to credit; the quid pro quo for access to credit being support for, or at least acquiescence in, bureaucratic policy goals. Once this notion is understood then such key elements of Japan's economic structure as main banks, *keiretsu* or corporate clusters, the convoy system for the nation's banks (*gososendan*), and lifetime employment fall into place. Most companies have 'main banks' that are expected

to support corporate activities irrespective of profitability. The MOF long administered a so-called convoy system that guaranteed the viability of all the banks. Corporate clusters anchored by cross-shareholdings allow each member unlimited access to group resources while imposing, on the other side of the coin, unlimited obligations to other group members. Companies are expected to provide for the life-long livelihood of core employees (*seisha-in*) (Murphy, 2000, p. 36).

Table 8.2 Features of Japanese and American decision-making

	Japan	*Traditional US model*
Goals and policies	Long-term growth and global vision, respect for people, sharing of common corporate philosophy and vision	Short-term profits, domestic market orientation, shareholder value, employees in excellent companies share a common vision
Information collection	Analytical (deliberate) centralised and participatory, flow of tacit knowledge	Analysis of profitability, centralised, only specialists have information
Creation of ideas	Incremental and innovative, emphasis on quality for customers	Aim at 'home run' (bold initiative)
Time horizon	Large investment for the future	Short-term profits
Evaluation	Consensus and risk-taking, generous about failure after due effort	Profitability evaluation profit responsibility of divisions
Interaction	Good interaction between departments, cooperation by virtue of mutual trust	Profit responsibility reduces cooperation between departments and promotes individualistic behaviour
Speed of implementation	Prompt after consensus reached	Prompt in respect of acquisition and divestment
Financial performance	Low ROI and ROE	High profitability
Products with world competitive strength	Watches, cars, cameras, televisions, CD players, LCDs, semiconductors, game machines, electrical components	Aircraft, computers, computer software

Under these conditions, not surprisingly, the process of decision-making follows a different rational from those found elsewhere. Hence the supremacy of the long-term perspective: the system provides for it. A corollary of this long-term orientation is that most Japanese firm have a system for long-term planning.

Information collection

Consensus is normal in Japanese decision-making, but in order to arrive at a consensus all the interested parties have to be provided with sufficient information, the collection of which can extend the decision-making process. Executive decision-making by one person might be quicker, but an advantage of decisions by consensus is that implementation is faster because the relevant information has been widely disseminated and understood. For instance the development of a new car by Japanese firms takes a much shorter time than that taken by American car manufacturers because the decision-making process involves the component suppliers and the marketing, production and development departments.

Idea generation – incremental and innovative

In this context incremental refers to new processes or patents being imported from the West and improved upon. Processes are improved and new ideas are generated through vehicles such as quality circles and suggestions systems. New models of cars and colour televisions are frequently introduced, in part because quality enhancement is considered important. Continuous improvements ensure that Japanese products have a strong reputation for quality in the world market. As Murphy (2000, p. 32) explains, 'The Japanese are obsessed with the relative standing on global markets of their manufactured products in terms of cost, quality and technological advancement. The profitability or price-earnings ratios of the manufacturers themselves have been, until very recently, almost irrelevant – even meaningless.'

Thus Japanese firms are highly innovative in respect of products and production processes. One thinks of, for instance, the crystal quartz watch, automatic cameras, transistor radios, mopeds, games machines, high-quality LCDs and DVDs: these are all Japanese inventions, as are the JIT and line-stop systems.

Time horizon

Japanese corporations enjoy a long time horizon because of the unique relationship they have with their banks. Most Japanese corporations engage in long- or medium-term planning and make large investments for the future, for example in research and development, employee training and the modernisation of equipment.

Evaluation – consensus and risk-taking

Evaluation is usually conducted by groups. The management committee makes strategic proposals and these are evaluated by committees and other groups. Group decisions tend to be analytical but not risk averse. Considerable information is collected and the risks analysed. Responsibilities are shared and positive opinions (rather than negative or carping comments) tend to be accorded high prestige (Clark, 1971; Gibson *et al.*, 1988). Not surprisingly the Japanese organisation is sometimes called a 'learning bureaucracy' because that is what it is: a highly structured organisation in which the virtue of learning is deeply embedded.

Integration

Considerable interaction between departments is normal and allows for the free exchange of information and ideas, which tends to produce better decisions. For example when Toyota developed the Prius (a hybrid car using both a petrol engine and an electric motor), many departments cooperated with the project development team. Good interaction is not left to chance, it is actively managed, while the lifetime employment system promotes mutual trust and excellent interpersonal communication.

Speed of implementation

As discussed above, once a decision has been reached, implementation tends to be fast because the relevant information has already been shared by all concerned and sense made of it. However the *divestment* of underperforming businesses is slow. American corporations such as GE are quick to close unprofitable departments and lay off redundant employees, but Japanese firms use slow and less direct means of cutting their workforce, such as reducing the number of new recruits, waiting for the retirement of older employees or transferring them to subsidiaries.

SURVEY OF LONG-TERM PLANNING

A survey was used to explore decision-making in Japanese firms. The survey was carried out in 1995, in the middle of the economic recession. Questionnaires were mailed to 200 manufacturing companies, of which 97 responded (an almost 50 per cent response rate). (See Appendix 2 for details of the companies.)

Table 8.3 Responses to the survey on long-range planning, 1995

	Response rate (%)
1. Have any changes been made to your long-range planning system?	
(a) The company has suspended long-range planning, and concentrates its efforts on short-range planning	7
(b) No change	48
(c) It has changed:	35
• Focus on strategic issue finding	
• Emphasis on profit improvement	
• Global development	
• Rolling or change of time span	
2. Features of present long-range planning system (overlapping responses)	
(a) Quantitative and forecasting	39
(b) Focus on strategic directions and policies	73
(c) Focus on strategic projects	15
(d) Strategic projects and quantitative integration	20
(e) Strategic issue finding	29
3. Planning process (some overlapping responses)	
(a) Top down, with strong leadership from the top	5
(b) Top down, planning department playing key role	15
(c) Interactive planning between top management and departments	69
(d) Consolidation of planning of departments	12
(e) Planning with subsidiaries and allied companies	11

Note:
1. Survey in 1995.
2. Responding companies were 97, questionnaires were sent to 200 manufacturing companies (response rate 50%). See Appendix 2 for responding companies.

Diffusion of a long-range planning system

About 93 per cent of the firms had a long-term planning system (Table 8.3). Japanese firms place considerable importance on such systems for strategic decision-making because their strategy is oriented towards the long term. Long-range planning is useful not only for systematically forecasting future trends but also as a technique for soliciting new ideas, creating consensus for decision-making and developing analysis of strategic issues.

A new trend is to use a top-down approach to identify future strategic issues. Strategic issues can then be explored rather than just extrapolating from past trends or just providing quantitative data that reports existing activities. Present features of long-term planning include exploring new directions for product-market strategy and deciding which strategic issues should be emphasised. Extrapolating from past trends is not so evident (about one third of respondents suggested this use). The focus is shifting from comprehensive integration to a project, future-oriented emphasis.

Mintzberg (1987, 1988) has suggested that formal long-range planning is dead and and is being replaced by 'emergent strategy'. Our findings suggest that, for Japan at least, formal and comprehensive long-range planning does not contradict emergent strategy. Such strategy is evaluated by planning. Quinn (1980b) follows Mintzberg in suggesting that the incremental approach works better; it is evident that Japanese firms use both the incremental and the comprehensive approach and do not regard them as conflicting but as complementary. Comprehensive planning is used to evaluate strategic options to allocate resources. In addition the development and production of, for instance, semiconductors, cars and drugs require a large investment, so analytical decision-making based on a considerable amount of pertinent information is needed.

The planning process starts with top management, which decides the basic strategy, and then the relevant departments draw up a functional or product plan that is discussed and approved by top management. Hence the process takes the shape of the letter U – descending from the top, through the ranks to the bottom, and then back to the top again – and is not the same as a simple bottom-up or top-down approach.

THE PART PLAYED BY JAPANESE FIRMS IN JAPAN'S ECONOMIC PROBLEMS

Since 1990 many financial institutions and service firms have experienced a serious financial loss and Japan's economic growth rate has been either static or in decline. Company failure has mostly been concentrated in the financial and service sectors and has rarely been seen seen in manufacturing. This section analyses the causes of failure and examines whether these include aspects of the Japanese management system. Several common hypotheses will be considered.

Japanese firms have been too growth-oriented

Many banks and other financial institutions loaned money for land investment through their subsidiary companies. Banks, particularly branch offices, were competing to increase sales and lent money to subsidiaries to avoid control by the MOF (Ministry of Finance).

Land prices underpinned loans to such an extent that many Japanese economists describe the Japanese system as a *tochi hon-i-sei*, or land-value economy. (Loans of more than one year were typically secured by land and most Japanese banks paid little attention to corporate cash flow.)

When the price of land started to decline after the burst of the bubble economy of the late 1980s bad debts accumulated, but the facts were not disclosed because there was no legal requirement for consolidated financial statements to be published. Japan lacks the institutional infrastructure of a modern capitalist economy in terms of independent ratings agencies, a well-developed accountancy profession and a business-oriented legal system (Murphy, 2000, p. 38) because in the past there was no need for them. Protection by the government meant that most financial institutions would not be allowed to go bankrupt, which only served to encourage a lax attitude by the managers of these institutions. Some small independent businesses, lesser corporate clusters and less important manufacturers were allowed to go to the wall, and some financial institutions were forced into 'shotgun mergers' by the Ministry of Finance, for example the Mitsui and Taiyo Banks were merged into the Sakura Bank. However these were mere window-dressing in politically weak sectors, oriented towards foreign perceptions rather than representing a substantial shift from standard practice.

In the past manufacturing was growth oriented and many producers now have a surplus of facilities and manpower. Nissan constructed a large plant in Kyushu and suffered a large loss because increased sales did not accompany the expansion. Renault has now taken over Nissan. Manufacturing firms rarely engaged in speculative investments because they were not protected against bankruptcy in the same way as the financial institutions were.

Japanese firms have been insufficiently analytical

Some of the companies which did invest did so without sufficient analysis. Yaohan was a supermarket company that made large investments in foreign countries, but some of these investments were in areas in which it had little knowledge, such as department stores. Toshoku (a trader in foodstuff) and Okura Shoji (general trading) were large companies with a long history but they went bankrupt because their subsidiary companies made bad investments in land. Many construction companies made large investments in land and experienced substantial losses. For instance the share price of Fujita (a large general construction company) fell from 2000 yen to 50 yen in 1998. Aoki Construction acquired the Westin Hotel chain and Seiyu (department stores) acquired the Intercontinental Hotel chain, only to sell them later because they had no experience in this line of business. In the majority of these cases the decision to invest was not arrived at by consensus but by autocratic executives (Fujita, Aoki and Seiyu) who acted on a hunch rather than basing their decision-making on informed analysis. Other cases were the result of decisions made by subsidiaries where top management had insufficient information.

Japanese firms were too oriented towards 'me-too' behaviour

Many steel companies entered into the production of semiconductors because it was a growth business that seemed to be related to their usual area of production, but they quit after making considerable losses, because they had no expertise in the product. Similarly the acquisition of five-star hotels and sizeable land investments were engaged in by one construction company after another: where one went others followed in a classic 'me-too' style.

Japanese firms were too conservative

Some firms did not take sufficiently aggressive action. While traditional clothing companies such as Renoun and Naigai were losing market share, others, such as Five Fox and Onward Kashiyama, increased their sales by launching fashionable new suits and introducing new supply chain management systems. So some firms were conservative, but this conservatism was hardly inherently Japanese. The camera manufacturers Yashika and Mamiya disappeared because they could not keep up with the speedy innovations made by other companies, such as Canon. Nissan, Mazda and Mitsubishi were slow to develop new models of car compared with Honda and Toyota and had to enter into alliances with foreign car manufacturers. Nissan's losses started in 1995 but the company failed to take drastic action to resolve the problem, and continued to announce profits by selling stocks and assets rather than engaging in innovative competition.

Japanese firms compounded their errors rather than take corrective action

When attempting to rectify their mistakes companies often made other errors and fell deeper into the hole. A good example of this was Yamaichi Securities, a large investment bank. By the 1990s its market share in the wholesale trade with business firms was falling and thus the price of the securities it bought for its customers fell. To hide the loss, shares were transferred to a dummy company at a high price. Yamaichi tried to recover the loss in the next round of trade but this also failed, thus the losses grew and the company went bankrupt in 1997. The errant trading was not the result of a consensus decision but was carried out by a small group of executives who were confident that the government would eventually step in to help the company. They had every reason to think this would happen because in 1965, in order to protect the public and to prevent a financial crisis, the company had been rescued by the Bank of Japan when its general stock price declined. Little did the management realise that, under US Treasury pressure, the rules were changing.

A similar change in the rules of the game was occurring in copper trading. A head of department at the Sumitomo Trading Company was a skilled trader in copper and top management trusted him because of his excellent record. However the price of copper declined in the 1990s when fibre optics began to replace copper wire. In order to recover the losses he had incurred with speculative forward purchases, he launched into further trade and ended up with a huge accumulated loss of about 200 billion yen in 1996 (representing the difference between the value of copper held by the company and the price at which it had been bought). In this instance the decision to buy had not been reached by consensus, as would have been the case in the manufacturing industry, but had been an individual decision by an appointed trader.

These cases of failure resulted from quite different decision-making practices from the typical Japanese practices shown in Table 8.2. Most failures were in the financial and service sectors and rarely took place in manufacturing.

Problems perceived to arise from Japanese personal characteristics

Some personal characteristics of the Japanese people have been judged by some as national weaknesses. In the following subsections we evaluate the truth or otherwise of such criticisms.

A tendency to say nothing

In university classrooms Japanese students rarely ask their lecturers questions, unlike students in the West. At business seminars Americans tend to express their opinions assertively but Japanese people tend to remain quiet. Even at company conferences proposals are sometimes approved without discussion. In the US, self-expression and competitive behaviour are necessary if peoples are to be recognised as outstanding, but Japanese people work in an atmosphere of equal treatment so they do not have to be as assertive.

Discussion In Japan, remaining quiet depends on the occasion. It is usual not to speak out in the following situations: at meetings with people from outside the organisation when the subject matter is not of serious concern to the company; in company meetings with superiors when the opinions expressed by these superiors are considered appropriate; and when consultations have been conducted before the meeting in order to seek opinions and factor them into the final decision, alleviating the need to air differences of opinion at the meeting itself.

However in other situations a considerable range of opinions may be expressed, for example during the development of a new product. If the opinions of the members of the project team and other specialists are not expressed, the project will not proceed. Japanese companies have developed many internationally successful products and if the project teams had been as quiet as the stereotype suggests, these products would not have appeared. When the subject is clear and ideas are requested, Japanese people will express their opinions. Japanese corporations have many committees and conferences where discussion is the norm, differences are displayed and decisions are arrived at by consensus.

Group decisions are considered the most effective form of decision-making in Japan and group members participate without reserve. The conditions under which divergent opinions may be presented are (1) when the subject or goal is clear; (2) when the participants come from different departments and have different areas of expertise, and hence are not in competition with each other; and (3) when the leader of the group solicits opinions from those present.

Group decision-making is slow

A frequent criticism of Japanese managers is that they do not make on-the-spot decisions when negotiating with another company, but refer the issue back to the head office. This is because decision-making is done by consensus and the head office is where the decision-making power is concentrated. Consensus decision-making manages power effectively by negotiation and protocol.

Discussion: There is no doubt that group decisions tend to take longer than individual decisions, but slow decision-making does not mean that decisions are made too late. Consensus decision-making has considerable merit, for example it tends to be analytical but risk-taking.

Japanese businessmen never say no

It is often suggested that Japanese people do not like to say no or to oppose other people's opinions. It should be appreciated that good interpersonal relations are important to the employees of Japanese companies because – rather like the fellows of a traditional university college – they can expect to be working together until retirement. They are afraid to jeopardise their friendly relations with colleagues and those who are senior to them.

Discussion: We recognise that Westerners feel free to disagree with others and that in most cases this will not destroy a relationship, but Japanese people carefully select the circumstances in which to be verbally negative. If the issue is deemed minor they will tend to refrain from commenting, but if the issue is serious they are perfectly capable of unequivocally expressing their disagreement. In failing companies it is very difficult for staff to oppose the top management, but in successful companies support and dissent will be voiced as appropriate.

CONTEMPORARY TRENDS

A review of the literature reveals the following general trends in strategic decision-making. Quinn (1980b) states that logical incrementalism is the most practical and easy to implement method of decision-making, (1988, 1994) while Mintzberg suggests that 'emergent strategy', coming from the grass roots or creative processes, is more important than strategy based on formal long-term planning. (Mintzberg, 1988, 1994d). Prahalad and Hamel (1990) point to the importance of core competencies, while D'Aveni (1994, 1995) advocates agile management (D'Aveni, 1994, 1995a). Clarke and Clegg (1998) suggest that strategic thinking is becoming more important than strategic planning. Some of these practices can be found in Japan (Table 8.4) but top management decisions remain final, strategic decisions remain more or less centralised and the concerted allocation of resources into growth areas remains important. (See Table 8.4.)

The varying degrees of initiative by top management have been reflected in considerable performance differences. In the period of high economic growth until 1990, companies were able to survive by following a me-too strategy, but this is not possible in today's conditions of slow or virtually no growth. However even in the slow growth economy there remain many opportunities. Technological innovation continues at firms such as Sony, and despite the recession companies such as Epson, Toyota and Ricoh have been very successful as a result of top management initiatives.

Top management initiative

With regard to the group decision-making style in successful companies, leadership by top management is particularly important. Top management initiative is evident in the concrete presentation of ideas, the ways in which policy decisions are made, the encouragement of data and information collection; through organising for innovation, and the evaluation of new ideas. Long-term planning replaces detailed planning. Company policy has developed corporate creeds that emphasise the initiatives of employees (see Chapters 3 and 10).

Aggressive information collection

Effective strategy involves consolidating head office competencies, especially in planning, research and development. Successful companies spend a large amount of resources on research and development and seek to develop and list patents – often in the US Patent Office. Some successful new products emerge from an explicit search for new opportunities. The Lexus was the result of Toyota's search for a flag ship car. *Utsurundesu* or 'Snapshot' film was developed by Fuji Film after studying the problems often encountered by amateur photographers. It was an investigation of the needs of consumers that led to the development of the 'Silent' Piano and a bicycle with an electric motor. Contemporary information technology makes communication with the consumer less expensive so that companies can collect considerable information and carry out one-to-one marketing.

Encouraging ideas from middle managers

Some successful new products have been developed from ideas put forward by middle managers. For example a young female system analyst invented the *Tamagochi*. She thought that an electronic toy that had to be 'fed' and taken care of in a variety of other ways in order to be kept 'alive' would be enjoyed by children. It became extremely popular and generated substantial profits for its originators, the Bandai Company. In order to motivate middle managers, specialist career paths and larger salary differentials are being introduced, as we shall discuss in Chapter 10.

Agile management

Agile management – the rapid grasping of new opportunities – is becoming a key success factor for firms such as Sony, Honda, Five Fox (clothing), and Epson (domestic electrical appliances).

Performance Evaluation

Profitability has come to play a more important role in performance evaluation, particularly in the case of current business. However, unprofitable businesses are not easily discontinued or divested, because employment has to be maintained. If a plant has to be closed, this is done over a period of two or three years and the employees are either transferred to other departments, or are offered early retirement or voluntary redundancy with an increased retirement allowance.

The evaluation of new businesses is based not only on profitability but also on estimates of future demand. Large investments in development and production, such as those required for large-memory semiconductors, are based on strategic evaluations. The use of calculations based on net present value and return on investment by the discounted cash flow method to evaluate new ventures is not popular among Japanese firms.

Established businesses rather than new ventures carry out the aggressive search for new opportunities in high technology areas because large investments are required. Project teams play an important role. Internal development, rather than the acquisition of other companies, remains the most popular method of new product development. Hence, initiative and creative ideas contributed by middle managers are encouraged and new personnel management systems introduced to recognise these skills.

SUMMARY

Strategic decisions focus on changes to product-market strategy or resource structure.

Strategic decision-making in successful Japanese corporations is long-term and growth oriented. There are two reasons for this. First, the company has an obligation to respect people in the internal labour market. Second, the structure of credit arrangements encourages growth in the asset base in the long-term. Decision-making by incremental search for consensus among various options, involving many committees, meetings and project teams, is normal. There is effective cooperation between departments because of lifetime employment. Decision tends to be slow but, after the decision is arrived at, implementation is fast.

The recession after 1990 affected manufacturing industries adversely. The failed companies decisions were: (1) too growth oriented; (2) insufficiently analytical; (3) me-too or conservative through the arbitrary exercise of top management power; (4) repeated initial mistakes in seeking to recover losses from earlier failed decisions and, thus, (5) delayed solution of critical problems.

Successful companies now place great emphasis on developing organisation initiative. Many companies use long range planning for identifying strategic issues. Group decision-making remains widely used and internal development, rather than acquisitions, is the normal path to new product lines.

9 Organisational Structure and Processes

Increasing speed of change and the emergence of mega-competition characterise the present competitive environment, producing new opportunities and threats. In order to cope with such a changing environment, organisations need several attributes (Figure 9.1). First, they need clear vision to direct change, orienting resource allocation to those niches where the company has the best chance of maintaining a competitive edge. Second, the company needs to be agile and flexible. Agile means responding quickly to new opportunities, such as the rapid development of new products. Flexibility requires frequent use of matrix organisa-

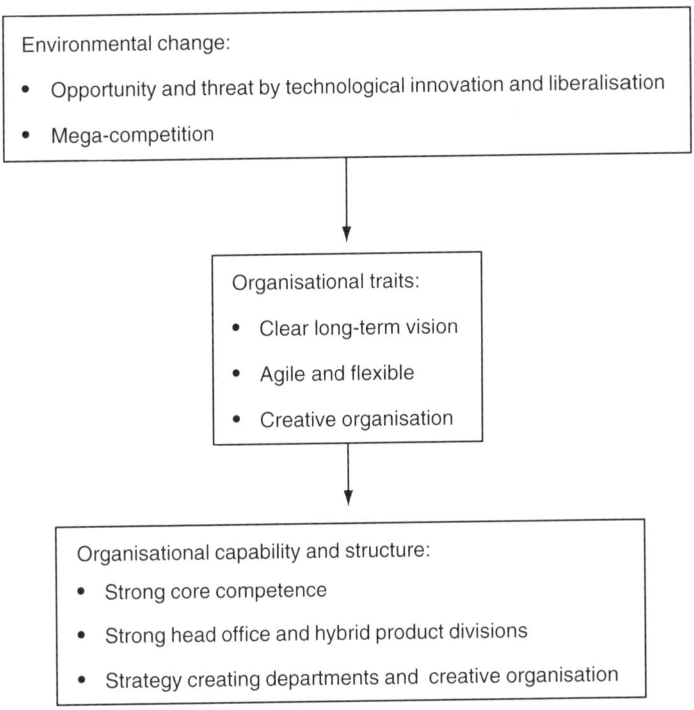

Figure 9.1 Organisational structure that can cope with environmental change

tions, team systems, small circle activities, good interface between departments and participation in decision-making by members of the lower echelons of the organisation. Third, the organisation needs to have strong core competencies. It is core competence that provides the capability for both producing successful new products and sustaining the competitive edge of existing products, according to researchers such as Penrose (1959), Ansoff (1965), Prahalad and Hamel (1990), and Hall (1993a).

CORE COMPETENCIES

Three characteristics of a core competence can be identified. First, it provides a common capability that can be applied across various products. For instance in NEC, electronics technology has been accumulated for many years. It has been applied to the production of communication equipment, computers and semiconductors, and has contributed considerably to the growth of the company. Canon possesses technologies for fine mechanical engineering, optics and electronics. These multiple core technologies have been applied to cameras, copiers, printer and other devices. Second, the capabilities that are embedded in the company must be needed by existing customers and be capable of producing high-growth new products. Sony's competence in producing small goods led to the creation of the Walkman, 8 mm VCR cameras, compact disk players and many other successful products. Third, the product must be difficult to imitate. For instance cutlery manufacturers in Tsubame and Sanjo sold their goods throughout the world but they lost market share to Korean competitors. In response they decided to develop new product lines that would be much harder to copy because they were an innovative result of the unique competencies they possessed as companies. They switched to much higher technology products, such as Titanium head golf clubs, components for space vehicles and high-quality kitchen goods. These products built on their strong competence in metallurgy, plating, casting and other metal processing technologies that were not imitable by others. The multiple skills of Seven Eleven Japan were also not imitable. However Yaohan, a large-scale supermarket operator, simply expanded its business globally into space-letting stores (New Jersey, Shanghai), a high-level department store (Beijing) and ordinary supermarkets (Japan). It did not have any core skills, and as its debts increased as the yen declined, it went bankrupt in 1998. Table 9.1 lists some examples of core competencies.

Table 9.1 Examples of core competencies

- *Seven Eleven Japan.* Finding fast-selling goods and slow-moving goods by POS information systems. About 1000 field counsellors transmit information personally between the head office and the franchisees. The head office recommends about new 1000 items a year to the franchisees.
- *Parcel express service by Yamato Transport.* 24-hour receiving service at convenience stores, information system using portable equipment, instant location identification of in-transport goods, next day delivery using 10 000 trucks.
- *Sony.* Sony's business creed – 'Sony explores the unknown' – signifies an innovative and creative corporate culture. Ability to produce small, high technology goods.
- *Canon.* Starting from optical and fine mechanical technology, it constructed electronics and systems technologies. It had the ability to gain the top share of the market by technological breakthroughs, starting as the number two entrant.
- *Komatsu.* Developed the technology to produce earth-moving machinery by benchmarking and surpassing Caterpillar.
- *Cutlery producers* in the Tsubame and Sanjo districts. Proficient in metallurgy, gilding and metal fabrication. They were able to switch from the production of knives and forks to the production of titanium-head golf clubs, components used in spacecraft and high technology kitchen goods.
- *Nintendo.* Production technology for game machines and organising a system of software suppliers.
- *Sharp.* Production technology for liquid crystal displays.
- *Sanyo Electric.* Following the business creed 'Earth friendly', production technology for solar cells and large-scale air conditioners.

If an organisation is to have a strong core competence it has to be created and maintained by the head office, to which we turn next.

STRONG HEAD OFFICES CREATE CORE COMPETENCIES

Forming a corporate strategy requires the development of the goals and product-market domain of the company. Another requirement is the integration and control of divisional strategies by top management, working with the planning, research and budget departments.

Expert functional staff departments, such as personnel, production technology and marketing, identify and develop appropriate sets of core competencies for the company. Their task is to accumulate knowledge and help line departments to develop competitive expertise in marketing and production, and to help create the knowledge base for these capabilities by investing in resources such as personnel, fixed assets and information systems.

The head office provides central services such as procurement, finance and computer support services, while the individual departments develop specialised expertise and carry out the related activities. Units that are not included in the head office are product division headquarters, research laboratories and sales offices, in other words those departments that conduct research and profit-making activities.

Strong headquarters boost competitiveness

The results of our head office surveys of Japanese manufacturing companies in 1982 and 1995 are set out in Table 9.2. There was a high ratio of personnel in divisional headquarters compared to personnel in the unconsolidated parent company, particularly in the more specialised companies. Between 1982 and 1995 that proportion increased.

Corporations that have related diversification tend to have a larger headquarters than those with unrelated diversification, because core competencies such as research and development capabilities are common across many products. So R&D and the search for new products tends to be centralised. On the other hand, companies with unrelated diversification do not need to centralise the search for new products, particularly when diversification is not carried out by internal development. Where diversification is carried out through the acquisition of other companies and the divestment of unprofitable businesses, the headquarters can be small. If the quasi-integration of suppliers and vendors is linked with several departments, their trade has to be dealt with by the central purchasing department or the marketing department at the head office. Also, such internal development needs the support of a strong core competence in research laboratories under the control of head office. Vertical integration and internal development strategies tend to produce larger head offices. Specialised companies also have larger head offices because of their functional organisational structure.

Japanese companies tend to have large head offices because they have a technology-related or a marketing-related diversification structure, so there are common competencies among their product groups. They extensively use quasi-vertical integration and strategic alliances, which also require strong head offices. Diversification tends to be by internal development and rarely by the acquisition of other companies. The

Table 9.2 Size of the head office: ratio of head office personnel to total employees[1]

Ratio over unconsolidated company employees (1982)			*Ratio over unconsolidated company employees (1995)*			*Ratio over consolidated company employees (1995)*	
Number of companies	*%*	*SD*[2] *(%)*	*Number of companies*	*%*	*SD*[2] *(%)*	*%*	*SD*[2] *(%)*
17 specialised	8.9	4.6	33 specialised	14	12	8	5
27 diversified	9.1	5.1	64 diversified	8	4	6	8

Notes:
1. The sample consisted of large manufacturing companies: 44 companies in 1982 and 97 companies in 1995 (mail questionnaire).
2. Standard deviation.

functions performed by the headquarters of successful companies are listed in Appendix 3.

Centralisation of decision-making

A number of principles underlie the centralisation of decision-making. First, under the division of labour the authority to take decisions tends to be delegated to those who have the necessary information and ability; typically those at the top. These upper-level managers possess strategic information on external, long-term environmental changes, which they use to select new markets, thus centralising strategic decisions. Lower-level managers are left to make decisions on operational issues. Thus, typically, the more innovative the company, the larger its head office and strategic planning departments. Small head offices tend to characterise less innovative companies (see Appendix 3). While no guarantee of success, because it depends on what is done at head office, we note that the more successful companies in Japan tend to have large head offices. There are exceptions of course, but these are organisations that are characterised by a lack of strategic competence, often due to monopoly or public sector control, such as the former Japan National Railways. Some companies with small head offices and large offices at the divisional level may achieve outstanding performance through tall organisational structures (such as the US firm GE).

Coordination of the activities of various lower-level departments requires not only rules and policies but also coordination of the activities at higher levels. Comprehensive planning and budgeting at higher levels are important means to this end. The mobilisation of key resources for important strategic projects is another key factor in innovation. There is a limit to the amount of information that can be handled. Head offices cannot handle too much information, so they must delegate authority. Delegation is also necessary for employee motivation because the centralisation of authority tends to demotivate staff at lower levels (March and Simon, 1958; Hall, 1982).

The advantages of large head offices

Companies with large head offices have sufficient resources to draw on to take innovative action. The application of massive resources to their semiconductor business by Toshiba, NEC and Hitachi are examples. Sony tripled its sales over a ten-year period by putting resources into electronics components, computer-related products (such as displays) and the expansion of AV products. The concentration of resources into these growth products would not have been possible under a decentralised product division structure. Strong corporate head offices also have the resources to create and sustain project teams in the incubator department of the head office for planning and implementing risky new ventures. A large head office also has more resources for the aggressive expansion of worldwide operations by establishing production centres around the world.

Effective head office organisation depends on the collection and management of corporate knowledge, data, information and know-how. Crucial questions for central knowledge management include future opportunities and threats to the domain. Such questions are typically not the responsibility of the product divisions; they often focus on more than two divisions, or can propose corporate restructuring in ways that the existing divisions would find difficult to address. Information on how to improve present products, improve quality and increase productivity is collected and used at lower levels. Diffusing it to other departments is a central role of the head offices.

If a company has decentralised product groups and a weak head office, the company can turn into a collection of separate small units that cannot easily develop an infrastructure of strong core competencies.

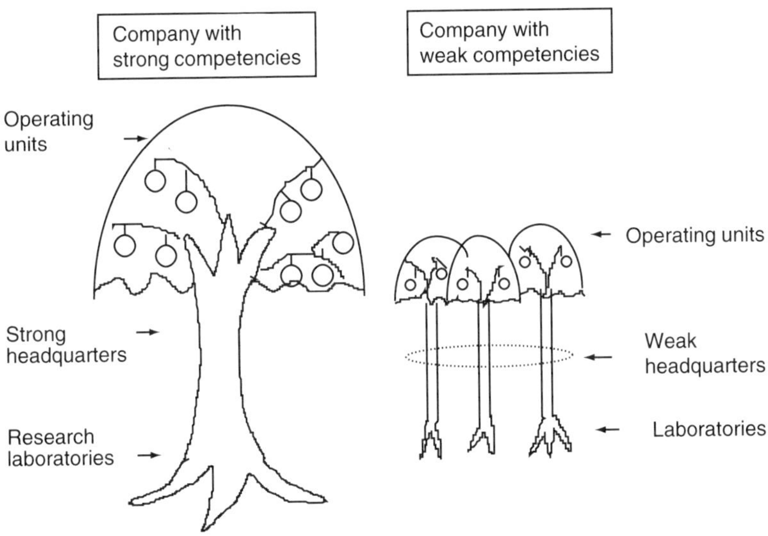

Figure 9.2 The head office as a source of core competence

In companies with strong head offices, central staff support is available for every functional area in all product groups. Such staff can accumulate knowledge from both external and internal sources to create and develop core competencies. This concept is illustrated in Figure 9.2.

The head office develops a personnel management system, a quality assurance system, production control technology and a marketing system. It also accumulates and transmits formal knowledge to line departments, making it possible to become 'a learning organisation'.

A strong head office also makes it possible to close certain departments and to change competencies – a product division is not likely to decide to abandon its own competencies. An important trait of learning organisations is the ability of head office to change competencies. For instance Hitachi and Toshiba moved away from heavy electrical and mechanical engineering and moved into electronics, followed by systems and software.

A strong head office makes it possible to maintain effective collaboration between product groups. At Canon a number of committees help to coordinate and control the activities of product divisions and foreign subsidiaries. These include the global operations committee, the global research activity committee, the global production committee, the global marketing committee, the trade balance problem committee and the environmental protection committee.

A strong head office enables a company to flatten its organisational structure. With a competent head office it is possible to make offices in the divisions simpler and this shortens the distance between top management and front-line employees. A company with a small head office does not necessarily have smaller overhead costs – less direct costs can create lower effectiveness with considerably higher indirect costs.

A large head office can also provide a pool of talented people. The head office of a reputable company is better able to attract competent people from outside and inside the organisation because of the prestige of the office and the quality of its jobs. This reserve of human resources can be used for the internal development of new businesses and for the establishment of operations overseas.

Disadvantages of large head offices

There are some typical problems with large head offices, including the 'over-control' of divisions and lower departments. A strong head office may concentrate authority and discourage initiative by product divisions. Also, divisional heads may tend to look up to and request help from the head office. Under the lifetime employment system a company cannot easily discharge an unsuccessful divisional manager, and this may result in less delegation of authority to the division, which in turn will cause the division to rely even more on the head office for decisions.

Elite staff at head office may not understand the actual situation at the front line, and in consequence may impose unrealistic controls on the divisions. Better communication, frequent visits from head office staff and job rotation can rectify this problem. Also, it is important that visiting head office staff are not treated as through they are on a royal tour, some rare and special event where only a rehearsed and sanitised version of the frontline reality is presented to 'white-gloved' and remote emissaries from on high.

HYBRID DIVISIONAL STRUCTURES

Companies with related-product diversification tend to have a hybrid organisational structure. The findings of our survey on the organisational structure of manufacturing companies are shown in Table 9.3. The survey shows that half of the diversified companies have a product-division, product-grouping or M-form structure and the remainder have hybrid structures. However a detailed analysis reveals that most of the product divisions are not fully fledged divisions because they lack some key divsional functions and have a strong head office.

Table 9.3 Divisional organisational structures

Product mix	*Specialised companies*	*Diversified companies*	*Total companies*
Product divisions integrating marketing and promotion	7	36	43
Hybrid, with two profit centres	5	13	18
Area divisions	2	0	2
Functional, partly product divisions for minor products	6	7	13
Functional divisions	12	7	19
Other	1	1	2
Total companies	33	64	97

Notes: Survey conducted in 1995 by T. Kono, by mail questionnaire and by reference to organisational charts. The total sample is 97 manufacturing companies. Specialised companies have more than 70 per cent of sales from one product line. These 97 companies are included in the 203 companies in Table 4.1.

Product divisions with a few functional departments

This kind of company employs a product-division structure but has strong head office support. The structure is shown in Figure 9.3a. About half of the product division structures in the sample

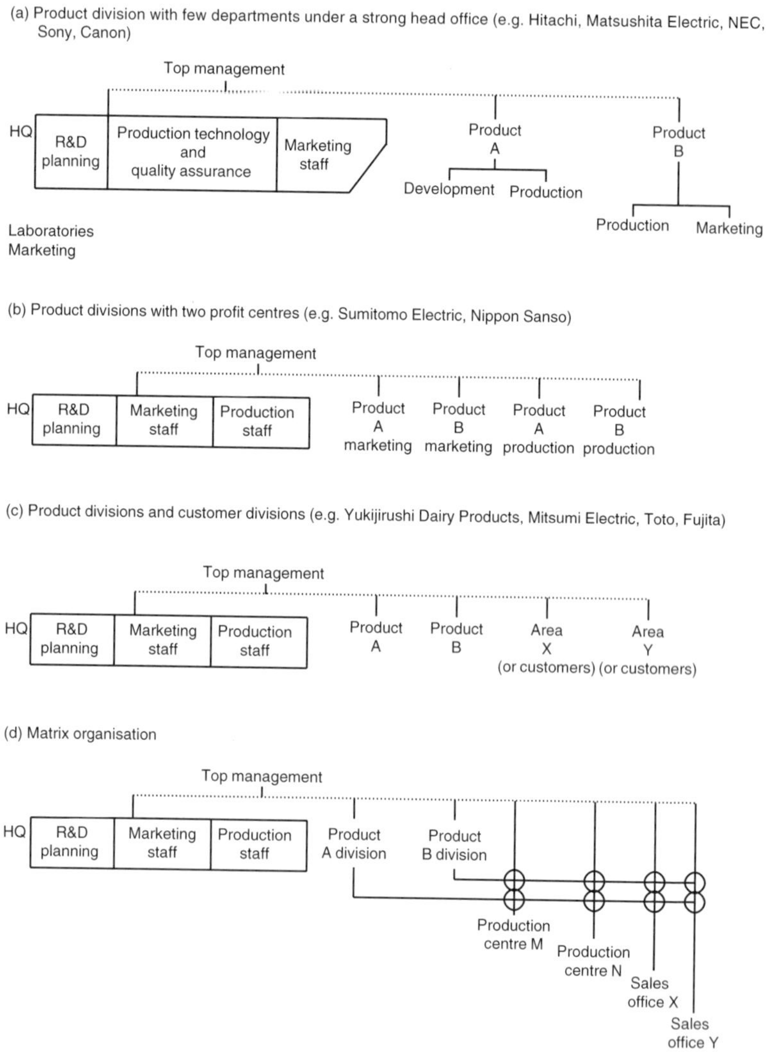

Figure 9.3 The hybrid product division

manufacturing companies belong to this type. The head office is strong. Research and development, production technology, quality assurance and marketing planning are centralised, and the research or marketing function are missing from product divisions of this type. Matsushita, Hitachi, Mitsubishi Electric and Sony all have this type of organisational structure.

One advantage of this structure is that the company can develop core competencies in research, production technology and marketing. Where there is related diversification of products, this centralisation of functions enables the company to maintain a strong competitive edge. So the advantages of this structure are that it has strong central functions, good collaboration between functions in the product divisions and a flatter organisational structure. A possible disadvantage is that it may discourage strategic initiatives by the product division. However this problem can be resolved if guidelines for a new strategy are shown to the product division and used as a vision to stimulate new initiatives.

Many new ideas come from researchers in centralised research laboratories. Toshio Ikeda at Fujitsu invented a new computer system in about 1970 that enabled Fujitsu to become one of the largest computer manufacturers in the world. The top management supported him, but not his divisional manager, who tended to be more interested in short-term profits.

Hybrid structure with two profit centres for each product

In this type of organisation each product has two profit centres – one in the marketing department and one in the production department (Figure 9.3b). The advantages of this structure are as follows. First, it has some of the benefits of both functional specialisation and integration. For example Sumitomo Electric Company produces electrical production equipment for various industries. The production and sales are carried out at different locations and need different skills. When the products are mutually related, the production department can develop functional core competencies. Second, the product profit centre system enables the departments to face market competition. Third, while this system is similar to that of functional departments, it is more flexible; moreover the mobilisation of resources to a project team or other strategic projects tends to be easier because the profit responsibility is not so strict. The transfer of personnel is easier than it is with the strict product division system. One problem with this organisational structure, however, is deciding transfer prices.

Product divisions and area or customer divisions

This structure (Figure 9.3c) is typically seen in the construction business and dairy product companies – one groups by products and the other by geographical area or customer. The merits of this structure are as follows. First, each output group has different knowledge, decision-making style and customer requirements. For example the production and marketing of milk and ice cream is different from that of other dairy products, however the local sales offices combine these products and sell them to the same customers. Another example comes from the construction industry. The technology for civil engineering and basic structures is very different from that for the construction of houses and interior design. However the customer can be provided with the results of a combination of these technologies. The area office has to identify and define the needs of customers, while the local sales office takes location-specific features into consideration.

Second, when products are mutually related in respect of production technology or marketing, this can have a synergistic effect on production or marketing. Third, this system is more flexible than the product-division type and is better for generating and developing new projects.

The matrix organisation

The matrix-type organisation is used where many products are produced, production is concentrated and sales are scattered (Figure 9.3d). For example Shiseido sells cosmetics and toiletries to a variety of consumers. Likewise Toto produces a variety of sanitary ware for households and offices, and here too the customers are scattered. The product managers take full responsibility for the marketing and production of each product, controlling the production centres and sales offices. The product manager is the only profit centre, and the production plants and sales offices are cost centres.

A similar type is also seen in the operation of subsidiaries of foreign companies. The local companies are responsible for profits but the product divisions in the home country have authority over production technology and the personnel management of the technical staff. Canon has a product-division structure and also a matrix structure. It has a number of committees with the authority to take decisions in several areas, as explained previously.

The merits of the matrix organisation are that it can integrate diverse parts of an organisation to try and achieve goals and allows specialisation, while new projects are easily organised and staff overlap can be avoided. A major problem is that employees have two bosses, which may cause a conflict of opinions.

THE DEBATE ON STRUCTURE AND PERFORMANCE

Organisational structure is mostly determined by (1) product or technological diversity and (2) market diversity. Using these two dimensions we can establish a typology for the line organisational structure. A pure product division is appropriate for unrelated diversification and a small head office. When the product is diversified but the technology or marketing is mutually related, then a product division with a large head office (that is, predominantly a product division) is appropriate. When products are diversified but mutually related, and when the customers or areas are diversified and strong responsiveness is required, then a two-profit-centre product division or a matrix organisational structure will be used. When the company's product is specialised and the customer or market is concentrated, a functional organisation is appropriate. Having elaborated this typology, it is worth noting that the above principles neglect the dynamic features of an organisation, such as the accumulation of core competencies and the dynamic mobilisation of resources to meet the needs of customers. They do not describe the functions of the head office and do not state the relationship between the line departments and the head office. The relationship between strategy and organisational structure is shown in Table 9.4.

It is frequently suggested that a large head office means high labour costs and reduced financial performance. However in Japanese manufacturing companies the percentage of average labour cost over sales is between 10 per cent and 15 per cent. The 8 per cent of total manpower employed at the head office may account for only 1 per cent of sales, so the head office personnel cost is not a serious burden to these companies.

Diversified companies can be classified into four groups according to whether the head office is large or small, and whether financial performance is good or bad. Some companies have large head offices and a good financial performance. NEC, Hitachi, Sony and Canon belong to this group. In these companies R&D expenditure is high, the products are mostly high technology, advertising expenditure is large (as at Shiseido) and the products are differentiated. The level ofdiversification is high but the products are related in terms of technology or marketing.

Table 9.4 Fitting the organisational structure to the corporate strategy

	Quality of movement	
Direction of movement	*Low technology or conservative culture*	*High technology or innovative culture*
Specialisation	• Small head office • Functional (sugar, cement)	• Large head office • Strong strategic planning department • Functional (Honda, Toyota, Marks & Spencer)
Diversification • Little movement • Large movement, acquisition and divestment	• Small head office	• Small head office • Product division
Internal development		• Large head office • Strong strategic planning department • Hybrid line organisational structure (Sony, Matsushita)

Some companies with a large head office have had a low average ROI (less than 5 per cent). These companies have a low R&D ratio, low product differentiation and little diversification, and are in low-growth industries. Their head offices do not produce aggressive strategies.

Other companies have a relatively small head office but their performance is better. These companies have a high R&D ratio, with a higher share of the market and less diversification. It is probably because of the latter that they can manage without a large head office.

Finally, there are many companies with small head offices and a low ROI. These companies have a low R&D ratio, little product differentiation, little diversification and make few changes to their product mix. These factors are due to the small head office making little effort to explore new business opportunities.

While the widespread use of IT has reduced the time spent on repetitive manual work and computation, it is not clear whether it has reduced the number of people employed at the head office or in other departments. Table 9.2 suggests that in Japanese companies the number of people employed at the head office has actually increased. The centralised economic planning and control system in communist countries failed because of the overcentralisation of authority. Is the same problem likely to arise in a strong head office system? There are many differences: in the head offices of successful private corporations, management takes strategic decisions and the majority of operating decisions are delegated to operating departments, motivated by the intrinsic and extrinsic rewards they achieve through successful competition.

Are head offices destined to remain remote in relation to innovation? Because of their remoteness, head offices do not always receive feedback on day-to-day experiences and experiments, and cannot accumulate knowledge through interaction with customers and production staff. Knowledge about strategy is produced in the research laboratory, the planning department and by marketing staff through experimentation, the collection of information and the reading of books and case studies. Knowledge about strategy is accumulated at higher levels in the hierarchy, while knowledge about operational improvements is obtained further down the line. It is often the case that head office staff are short of information about the situation at the front line and they may put forward ideas that are too abstract or out of date. For example the personnel departments of many Japanese corporations introduced an American-style wages system after the Second World War, introducing job-related pay, job evaluation and the classification of jobs. However the system did not work because it was not congruent with traditional Japanese values, where pay for skill was seen as fair and pay was expected to reflect length of service and cost of living. The pay system was later changed to one that integrated pay for skill and length of service.

It is sometimes suggested that head office staff tend to be conservative because they are specialists, housed in comfortable headquarters, enjoying great prestige but with a tendency to use their remoteness from the actuality of the organisation to overexert their power. The criticism that it is conservative ignores the role the head office plays as a research laboratory for management. Head office staff may have this sort of tendency but there are ways of avoiding the problem that Japanese corporations use routinely. First, job rotation: staff at the head office are usually required to spend time working at the front-lines in order to obtain knowledge of the company's operations and understand the feelings of front-line employees. Staff at head office are also recruited from line departments and are rotated to other departments. Second, they are encouraged to have an external orientation. Head office staff visit the front-line to make observations and they collect external information through formal or informal networks. Benchmarking is used to investigate the best practices of other companies. Thus 'management by walking around' and an externally oriented culture are developed. Third, a service orientation is encouraged among head office staff. The head office is required to be both a service department and one that produces a proactive strategy for the company. The head offices of successful companies play the role of strategist and act as a management research laboratory for top management.

Employee participation in small groups is essential for the creativity and success of a firm, but a large head office tends to create bureaucratic systems and bureaucratically minded people. The head office has three functions: to create new strategies, to provide expert services, and to concentrate service expertise in one place. It is important that the attitude of the head office should be innovative and creative in searching for new systems. This is the opposite of a bureaucratic attitude, and successful companies such as Honda, Sony and Canon, all with large head offices, have carried out continuous innovation and used participation to achieve it. The participation of employees in decision-making is possible at the head office, in the research laboratory and in lower-level departments, making participative management possible in a centralised organisational structure.

While the leadership and vision of top management may be more important than that of bureaucratic head office staff, successful top management makes full use of the staff at head office. Konosuke Matsushita at Matsushita Electric and Mitarai and Kaku at Canon created powerful research laboratories and strong head offices, and they listened to information and ideas from subordinates. They made decisions based on the information provided by their staff and engaged in long-term planning. Their leadership and vision were based on the support of the staff at the head office.

External communication is changing the boundaries of the organisation. As strategic alliances increase, intercompany communication expands in the interest of cooperation and integration. Many employees are sent to foreign countries to visit affiliated companies, as well as being required to communicate with domestic partners. This kind of integration needs people whose authority extends across the boundaries of the organisation. Also, interaction between departments within the organisation becomes more important. This integrating activity can be carried out at the divisional level, but to a large extent it should be carried out at the head office level because it is an activity that covers many areas. Many Japanese companies situate their marketing departments in the head office because they need to coordinate the activities of suppliers and distributors that deal with products from several product divisions. Also, the production technology department at the head office negotiates with suppliers and transfers quality control systems and JIT (just-in-time) systems to these partners. This is not to say that the head office carries out all the activities involved in the external customer–supplier partnership, but it plays a significant role in these activities. At the Bridgestone Rubber and Tire Company each plant has been linked to a foreign plant formerly owned by Firestone and is responsible for improving the production system and exchanging information. However strategic decisions on multinational management are taken at the head office.

Interface between departments, effected through project teams and the rotation of personnel, is more important when products contain a number of high technologies, and is made easier by lifetime employment. It is also made easier by the use of a hybrid divisional structure.

Network teams are more effective than formal hierarchies. The group system has several features. Groups change from time to time, depending on the needs of the organisation. One person may belong to more than one group, so flexibility increases and the organisation is made flatter because posts such as assistant head of department, section manager and subsection manager (*kakaricho*) are eliminated. Both Toyota and Nippon Steel have discovered the advantages of flexible groups. New assignments and the status ladder system can mitigate the effects of a decrease in the number of promotion posts. The use of groups can reduce the number of employees at the head office and increase its flexibility.

ORGANISATIONAL STRUCTURE AND STRATEGY

Strategy is one of the key functions of the head office. Under the head office, the research laboratory investigates technological breakthroughs in respect of new products and production processes. Japanese corporations have many powerful departments that deal with strategy.

The corporate planning department

There are many planning sections in operating departments below the head office. Here we shall concentrate on the central planning department.

The corporate planning department, sometimes called the president's office or general administration department, engages in the study and implementation of corporate strategy. The department has five important roles. First, it promotes strategic thinking. It signifies the importance of corporate strategy for top and middle management. Managers tend to focus more on daily operations than on strategic problems because the former involve less uncertainty than dealing with things that will bear fruit only in the long term or are inherently risky.

Second, the corporate planning department collects strategic information. It assesses the general environment, looks for new business opportunities and makes industrywide predictions. It also conducts internal analyses of the firm. Some of its activities overlap those of individual departments, which conduct analyses of the branch of industry in question and their most important competitors.

Third, the department has a planning function: it determines basic goals and policies and presents innovative ideas on strategy for the organisation as a whole and sometimes for individual divisions. In companies that employ the growth-share matrix model, the corporate planning department tends to have centralised authority. Also, new business that does not inherently belong to any existing operating units has to be sought out by the corporate planning department.

Fourth, there is the coordination and integration of the strategic plans of the operating units. One means of coordination is resource allocation through a long-term profit plan and a personnel plan. The fifth role is to check up on the implementation of plans by comparing the actual results with the milestones set for the projects. The department also assesses the quantitative performance of plans. While actual follow-up and performance appraisal are part of top management's task the planning department prepares the relevant information. It also draws up the standards to be used in the appraisal.

If the planning department's duties have a heavy emphasis on the collection of strategic information, the department may be classified as a research-type office. If the duties mainly involve the preparation of corporate strategy, the department may be classified as a strategic-planning unit, or a 'general staff office' in military terms. Many Japanese planning departments are of this type. If there is more emphasis on coordination, the planning department can be called a coordination office. The planning departments of diversified American corporations are mostly of this type.

According to our 1989 survey of 249 large companies, the average number of personnel in planning departments is approximately 7.5. If assistants such as female secretaries are excluded, the average number is 5.5, two of whom are typically natural science graduates and the rest are social science graduates. In the electricity and gas supply industry the number of personnel in the planning department is very high: 55 on average.

Project teams

Project teams are formed at the corporate, product division and research and development levels. They study strategic projects, new product development and other current issues, and sometimes implement measures. Team members have different capabilities and work towards clear goals, with set schedules and within a predetermined budget. When their project is completed it is delivered to the relevant department and the team is dissolved. As the degree of innovation and the number of new projects increase, so the use of project teams increases.

Teams can be full-time or part-time. Full-time teams are more powerful and are usually set up to deal with large, important projects. The responsibilities of the individual members are clearly defined, their work is never interrupted and good teamwork and communication are expected. However some departments are reluctant to release competent people to work with a team as this might damage the performance of the department. In our 1985 survey on new product development, about 50 per cent of teams were full-time and 50 per cent were part-time.

Project teams are used far more in Japanese corporations than in American and British corporations. There are two reasons for this. First, Japanese corporations have a relatively centralised authority structure. The profit responsibilities of product divisions or line departments are not very strict, so they are usually willing to send competent personnel to serve on project teams. Second, because of the lifetime employment system people are more willing to be moved around as employment is assured and promotion to a higher status grade can be expected as a result of serving on a team.

The success factors in project teams were also surveyed. Success in this respect means that the team members work hard day and night, look at problems from new angles and create successful new products, production systems or strategies. The development of the crystal quartz watch at Seiko, the Walkman at Sony and the word-processor at Fujitsu are all examples of successful new product development by project teams. The following factors seem to be important to success.

First, it is necessary to gain the support of top and middle management in order to facilitate cooperation with the relevant departments. A capable project leader is important in this regard. Moreover teams need to collect a considerable amount of information so the cooperation of other departments is necessary. Second, the teams' goals, responsibilities and schedules need to be clearly defined – clear goals are necessary to motivate team members. Third, it is important to have outstanding project leaders and high-quality team members with complementary capabilities and knowledge (For further information on success factors see Peters and Waterman, 1982; Kono, 1987).

The Sharp Corporation has launched many successful new products in recent years, including LCDs (liquid crystal displays), an 8 mm video camera (1992) and a portable electronic notebook. They were all created by '*Kinpro*' or emergency project teams over a period of six to eighteen months under the direct control of the president. The teams were composed of the firm's most competent personnel and were entitled to wear a golden badge, the same as that worn by directors.

Internal venture teams

An internal venture team is similar to a project team but differs in two respects: it is set up by an individual who has conceived a new idea, not by the corporation, and the venture leader becomes the manager of the project and takes a share of the profits (as in the 3M model). At present internal venture teams are not used very much by Japanese corporations because there is a shortage of capable development engineers, particularly electronics engineers, so companies cannot afford to release them from their normal duties to pursue what may be risky ideas. Nonetheless the system is widely recognised as a key means of motivating engineers and encouraging creativity.

Incubator departments

Many companies have established 'incubator' departments to develop new products. A case in point is the T Company, which produces tapes for music and for other purposes. The company has five research laboratories for new product development. Once a new product has been developed it is transferred to a 'future product department', an independent department that, in cooperation with related departments, conducts experiments on production methods, process improvement and marketing. When a new product is given the go-ahead it is either transferred to an existing product division or a new product division is established. Toray, which produces synthetic fibres, also has 'new

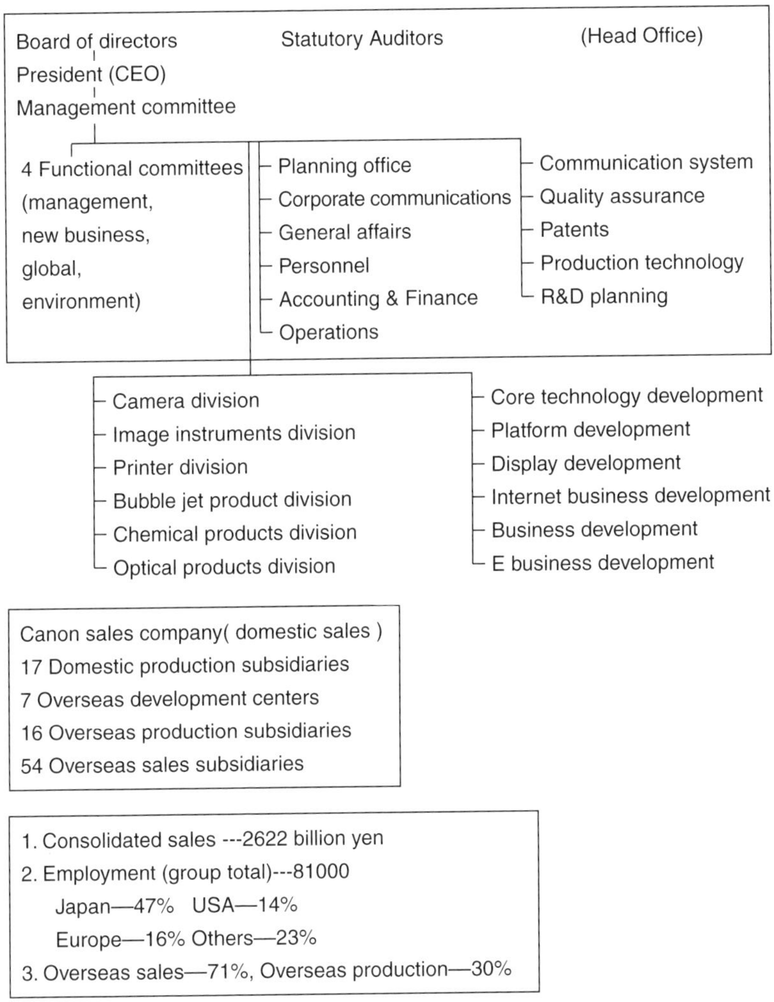

Figure 9.4 Canon Company organization chart (as of 1 April 2000)

business departments', where new pharmaceutical products, Mictron-Para-based Aramid ultra thin film and contact lenses are developed, in cooperation with the research, production and marketing departments.

Permanent product divisions tend to be more interested in short-term profits than in the development of new products that will not produce immediate profits. It is for this reason that incubator departments are established. Sometimes a subsidiary company is established instead, however such companies tend to receive less support from other departments than do internal departments.

Research laboratories

Research organisations are usually divided into laboratories for central research, new product development and production technology research. Most Japanese companies have centralised laboratories under the control of the head office. American and British corporations tend to have decentralised research laboratories under the control of the product divisions, although the central research laboratory is under the control of the head office. This centralisation makes it possible to concentrate research resources into new strategic projects to adapt to large changes in the environment. An advantage of decentralisation is that it eases the interface between new product development, production engineering, production and marketing.

Cooperation between research, development, production engineering, production and marketing affects not only the research results but also the speed of development, quality and cost of new products. For example the development time for a new car in Japan is one to three years while in the US it is two to four years. These differences in lead time can be explained by a number of factors, chief among which are systematic interdepartmental cooperation and long working hours. New product introduction tends to be more frequent in Japan (new models of some domestic electrical appliances are introduced every six months) and the vertical integration (or quasi-vertical integration) of components suppliers and sales channels ensures that there is a sound cooperative basis for new product development.

Good interdepartmental cooperation depends on good communications. There are several means of ensuring this. First, the high status accorded to production engineers facilitates cooperation between the research, new product development and production engineering departments, so it is important for the firm to have a relatively strong production engineering laboratory with many researchers. For instance Honda's production engineering department employs about 2400 researchers and other workers to study production processes (in total the research and engineering department has about 7700 employees). Second, there has to be broad participation by the production and marketing departments. Third, there should be a widespread exchange of personnel and technology transfer. Fourth, use should be made of parallel development. Several concepts will compete with each other with one concept eventually being selected. Other types of parallel development occur between stages in development. Research, development and production engineering schedules overlap at both ends of the process. Fifth, the project teams should be drawn from the development, production engineering, production control and marketing departments, as exemplified by Honda's DPS team (development, production, and sales). Sixth, there should be frequent meetings and researchers should regularly visit the plant, sales offices and customers.

Locating the research laboratory at the plant site is sometimes used to improve communication, but it is not a common practice. Research laboratories tend to be rather centralised, as already mentioned. Group orientation, a fundamental Japanese characteristic, ensures that mutual cooperation is the norm, and the fact that researchers, development engineers and production engineers are of relatively equal status contributes greatly to this cooperation. Moreover strong support flows from top management, whose involvement in research is extensive. About half of top management have scientific backgrounds and thus are able to understand technological issues. In addition, employees are very aware of the intensely competitive environment in which their firms are situated and understand the need to cooperate with each other so that the firm can withstand the strong market pressure.

THE CREATIVE ORGANISATION

Japanese research organisations have created many successful high-tech products that now dominate the world market. In this section we explore the factors that have contributed to this creativity. Table 9.5 lists the key factors and the characteristics of creative individuals (see also Steiner, 1965; Nystrom, 1979; West and Farr, 1990).

Table 9.5 Characteristics of creative individuals and organisations

Creative individuals	*Creative organisations*	*Less creative organisations*
Goals		
Like to challenge complex problems	Corporate creed emphasises creativity and long-term goals, allows freedom within clear strategic goals. Integration of individual interest with the goals of organisation	Constraining culture concentrates on short-term profits. Fearful of risk. Too much freedom or overly strong control. No integration
Information		
Rich knowledge, problem-oriented information collection	Distinct knowledge of selected basic technologies. Extensive collection of outside information through personal contact with outside sources. Exchange of diverse knowledge and information through free channels of communication	No distinct knowledge. Information internally oriented, emphasis on production technology. Poor communication between departments
Ideas		
Look at things from new angles, generate unusual ideas	Welcome new ideas and free atmosphere. System to collect ideas. Employees encouraged to challenge the impossible	Dismissal of new ideas or resistance to them if they emerge. Emphasis on operational matters and realism
Human relations		
Independent, nonconformist, see themselves as different	Respect for individualism. Mixture of top runners and later entrants. Respect for minority opinions but seek consensus. Research and development laboratory is separated from operations, but communication between them is opened as needed	Little respect for individualism. Always followers. Too much emphasis on consensus – the nail that sticks out is hammered down. Research laboratory is an ivory tower, or too preoccupied with requests from operational departments

Table 9.5 *Continued*

Creative individuals	*Creative organisations*	*Less creative organisations*
Evaluation and development		
Conduct inquisitive, long-term explorations. Patient. Look for refined solutions	Ideas are appropriately evaluated and developed. Research: balance between inquisitive exploration and speed, considerable slack	Ideas are not developed. Research: emphasis on efficiency, no slack
Motivation		
Motivated by interest in problem	Satisfy the need for recognition and achievement. Work-assignment: integration of individual interest with the subject of research, freedom to choose means of research. Respect that researchers want to be recognised outside the organization, freedom given to publish research papers, loyalty allowed to professional or academic organisations. Dual promotion ladder, selection and promotion by merit only, rewards for creativity. Freedom to fail, evaluation by additions, no deductions.	Little satisfaction of need for recognition. Work-assignment: assignment decided by the supervisor, strong control of research, or too much freedom. Emphasis on short-term performance. Loyalty to the organisation is stressed. Promotion by administrative capability only, too much emphasis on equality and length of service, or punishment is used. Evaluation by deduction only

Our survey on research management investigated the characteristics of creative people (Kono, 1992b). Individuals' responses to unstructured questions suggest that creative people like to challenge complex problems and possess knowledge that is both highly specialised and wide-ranging. They are also able to look at things from new angles, have a strong intuitive ability and strong faith in their own judgement and opinions, and are patient in the pursuit of answers to problems.

We also sought to identify the features of creative research organisations, and to compare these with the characteristics of creative people. Sanyo Electric provides a good example, (see also Kuwano and Yoshida, 1997). Sanyo's corporate creed – 'Create more comfortable environments and enrich people's lives' – emphasises creativity and tries to integrate individual interests with the goals of the organisation. Application of the creed can be seen in the development of products such as the solar cell, in which product Sanyo has considerable expertise and sales. Sanyo spend 80 billion yen, or about 8 per cent of its sales receipts, on R&D and about 6000 researchers (out of a total workforce of about 30 000) are engaged in R&D.

In order to integrate the individual's interest with the goals of organisation, research topics are decided in a top-down and bottom-up process. Such a system enables researchers to apply to develop research subjects that interest them. Sanyo encourages the development of distinctive knowledge and good communication between departments. Researchers are encouraged to attend academic meetings and the firm gives awards to outstanding papers that appear in journals. Researchers can apply for overseas study. The company has a career development programme for researchers: 'When they are about 27 years old researchers will present papers at domestic academic meetings and write a paper in English; when they are about 30 years old they will present papers at international academic meetings, and when they are about 33 years old they will gain a doctor's degree.' The company has an 'RD and B Team', with members from the research, development and business divisions, whose task is to integrate the three departments' knowledge and create successful new products. The rotation of researchers between the research, development and product departments helps this kind of cooperative research activity. Looking at things from new angles and producing a new combination of known elements is encouraged. Such a creative process allows for changes in the direction of analysis, and if one trial is not successful another will be set up, based on a fresh approach. Looking at things from new angles is a key process. A well-known training thought-experiment is to connect nine square points with four straight connecting lines, which requires one to expand one's imagination.

Management processes that encourage new trials include soliciting research subject ideas from researchers, allowing bootleg or under-the-table research, and forming research teams with an eye to creating diversity among the members. Sanyo's research laboratories encourage bootleg research of two types: 'free research', with the permission of a sector head and using the reserve funds of the section, and 'hidden research' during or after office hours (although not all laboratories allow the 'hidden' option).

Creativity can be promoted by assembling a group of people with diverse knowledge and approaches. New products are frequently developed by setting up a team that combines researchers, production engineers and marketeers. Combining people with a long-term approach with others with a short-term approach, or an inductive with a deductive approach, are cases in point. Other combinations include the blending of concrete skills with abstract skills, imaginative approaches with incremental approaches and so on.

However differences in values between the team members may be a problem, for example some researchers might place less value on promotion within the company and more on writing papers and obtaining prestige in academic circles. There are two opinions on the workability of teams where divergent values occur. One suggests that divergence is not desirable because gaining cooperation in research is difficult enough anyway, while the other suggests that when values are different, there is less competitive conflict among members so cooperation will be better.

Differences in ability is another issue. According to our interviews with research managers, basic research is best conducted by a group of first-class researchers because they stimulate creative ideas in each other. But at the development stage, when the goals are clear, better outcomes are produced from a mixture of first- and second-class researchers because control is more important (on this subject see Pelz and Andrews, 1966).

Respect for individual uniqueness and good cooperation between colleagues and between the laboratory and the operating departments are important. Creative people tend to be independent and nonconformist. However cooperation is required for present-day creative activities because high-technology research requires a variety of high-level knowledge that one person cannot possess. Independence and cooperation are opposing features – the former is a primary need of researchers while the latter is a primary need of the organisation (and sometimes one of the needs of researchers) – that have to be integrated if the research organisation is to be successful. Even in universities complete freedom does not necessarily lead to success (the development of what became the Macintosh technology by a researcher at Xerox's Palo Alto Research Cooperation facility is a case in point: viewed from Xerox's perspective this was not successful research).

Harmonising individuals' desire for independence and the necessity of cooperation is a key problem. One approach is to solicit ideas on research from the researchers and to evaluate them according to the corporate strategy. Another is to allow and encourage academic activities, as at Sanyo. As discussed above, Sanyo rotates its research workers from basic research to development to project teams to product divisions. In this way researchers not only broadern their knowledge, they also form relationships with personnel in other departments.

Creative research seeks that which seems impossible to find. Thus patience by both the researcher and the organisation is required. It took twelve years (from 1968–80) to develop the silicon solar cell at Sanyo. During that period the lead researcher wanted to abandon the project because of the difficulty of finding the elusive new material. However the general manager of the central research department decided to continue the research, which eventually led to Sanyo gaining considerable strength in the field of solar cell systems for electricity production. It is not uncommon for a research project to take up to 10 years, thus tenacity among researchers and the encouragement of senior management are essential. The long-term view adopted by Japanese corporations is advantageous in this respect.

Creative researchers need their achievements to be recognised not only by the company but also by the academic world, hence Sanyo's encouragement of researchers to write academic papers for outside journals, make presentations at academic conferences and obtain doctor's degrees. Researchers are evaluated not only in terms of economic achievement but also in terms of academic achievement, which is reflected in promotion. At Sanyo there is a prize for outstanding papers. Those whose discoveries are patented receive 2 per cent of the royalties, which can be as much as two million yen a year. For those whose research is not patentable because of confidentiality or was part of a group activity, other rewards are offered (on research management in Japanese corporations, see Kono, 1988; Okamoto, 1991; Kuwano and Yoshida, 1997).

CONTEMPORARY TRENDS

Intracompany units

Large manufacturing companies such as Sony, Toshiba, Hitachi and others have established intracompany units. These sometimes correspond to the product groups or sectors of American corporations in their degree of decentralisation of authority – mid way between highly and barely centralised. The units are not legally independent companies but departments within the company or a collection of product divisions. Usually the units have production, marketing and research functions. In the past, plant managers at Hitachi were responsible for profits and hesitated to make large investments because, if these failed, they would be reprimanded for incurring a loss. For this reason they did not take the number one share of their product markets, LCDs being a case in point. The new internal units have more power and authority to allocate resources, and Hitachi expects them to be more agile and to allocate resources rapidly to growth areas.

A problem with this system is the overlapping of functions among units. Thus Sony reduced the number of units from ten to four and combined their marketing and research functions. Sony also bought back the shares of its successful subsidiaries, including Sony Music Entertainment and its insurance companies, in order fully to control them. In this way it could move resources to growth areas in accordance with top management strategy. There is a tendency for subsidiaries that are losing money to rely on the parent company, whereas subsidiaries that make a profit try to distance themselves from the parent company and refuse to move their resources to other departments.

Replacing departments with groups

Many companies have abandoned departments in favour of groups with flexible responsibilities. Some employees belong to more than one group, working under two or more group leaders. A planning group with 10 to 30 members is responsible for numerous activities, such as long-term planning, product group planning, new project planning and support, foreign direct investment planning and support, company acquisition and alliance planning, long-term forecasting and the provision of clerical support for the management committee. Both group members and duties change as needs change. When the group system is applied throughout the organisation the hierarchy is flattened, for example two departmental and sectional levels are reduced to just one level of groups, reducing the number of staff and changing the nature of their jobs. The group system differs from self-management units such as quality control circles. There is a clear responsibility and line authority system. Positions such as department manager or section manager disappear, to be replaced by group leader. However the status ladder remains, thus the promotion ladder still exists.

Finance and strategy

Annual accounts reported unconsolidated financial statements against which income tax was charged. But this system is changing. Under the old system the parent company could boost its apparent sales figures and hide its losses, sell goods at a high price, and sell overstocked goods to its subsidiaries. It could also transfer its excess staff to its subsidiaries. These practices are changing. Parent companies are less able to use their subsidiaries to window-dress profits and are ceasing to protect these subsidiaries. It is becoming increasingly common to sell unprofitable subsidiaries and acquire profit-making companies, a strategy that does not sit well with the *gemeinschaft* principle.

SUMMARY

The head office has three main functions. First, to develop new strategies and help top management through, for example, the strategic planning and new product development departments. A strong head office can help to produce an aggressive product-market strategy. Second, to accumulate specialised expertise and help line departments. Strong expert staff can strengthen the core competence of a company. Third, to concentrate specialised activities in the finance, procurement, patent and information processing departments in order to build core competencies.

A strong head office will have the ability to create competitive advantages for the company. The capability needs of the head office vary with a number of factors. Associated with high capability needs are factors such as high technology products, diversified products related by technology or marketing, growth and diversification achieved by internal development rather than acquisition, extensive use of vertical integration, and a high level of foreign direct investment.

Japanese corporations have long tended to have large head offices, employing about 8 per cent of the total workforce. A large head office can initiate strategies with long lead times, something that product divisions cannot contemplate because such initiatives do not produce immediate profits. Also, the head office can commission projects that require a large investment, such as the development of semiconductors, or knowledge from two or more divisions, such as a combination of software and hardware, or electronics and mechanics. These projects may take place in laboratories directly under head office control, or be undertaken by new project teams either at the head office or in a new department, or by coordinating several product divisions.

A head office functions in a similar manner to a large management laboratory, accumulating knowledge on personnel management, production systems, quality control and marketing systems. Hence it is in a position to support product divisions and other operational units. The strength of companies such as Hitachi, Matsushita, Sony and Canon derives from such knowledge accumulation and strategies. If product divisions are separated and the head office is weak, then the entire organisation will be weak. A strong head office can sustain a flat organisational structure. Typically, companies with small head offices have large divisional-level offices. Strong head offices provide a pool of competent personnel to support foreign investment and new projects. Strong head offices also support the majority of the product divisions. The product divisions have no research laboratories, no control of the sales channels and rely heavily on head office for personnel management, procurement, production control and marketing. We call this an imperfect product structure. The hybrid organisational structure refers to divisions with profit centres. Imperfect product structures and hybrid product divisions fit well with technology- and marketing-related diversification, which are popular among Japanese corporations. They tend to be flexible and offer strong core competencies.

A contemporary trend is to moderate the centralisation of authority by establishing intracompany units and replacing departments with flexible groups. The central planning and research departments, project teams and 'incubator departments' play important roles in the development of successful strategies. The creativity of these departments, particularly the research and development department, is a key factor in the successful development of new products and processes.

Key strategic principles include long-term vision, the integration of individual interests with the goals of the organisation and the recruitment of high-calibre people with diverse knowledge to support the strategy. This pays off in constant, high-quality product innovation and the lodging of patents at the US patent office, and is fed by a steady, high-quality supply of science graduates from Japanese universities.

10 Personnel Management

PERSONNEL MANAGEMENT SYSTEMS

The personnel management systems at successful Japanese corporations have often been explained in terms of *gemeinschaft* (community organisation), as opposed to *gesellschaft* (association). Tönnies first proposed these concepts in 1887. A *gesellschaft* organisation is a purely profit-making organisation. Unless they are rewarded, people will not work. No moral or spiritual unity can be presumed. People are bound together by contract but they exist apart from each other, alienated and in a state of tension. They work according to a division of labour, within the strict limits of the job, and each is merely one small cog in the organisational wheel. A *gemeinschaft* organisation, on the other hand, is more like a family or a church, where the members are held together by bonds of love. In such an organisation working together is seen as a source of joy. People empathise with, and help, trust and understand one another, sharing bad luck as well as good.

The *gemeinschaft* concept may seem appropriate for understanding Japanese organisations, but it is too simplistic. *Gemeinschaft* may describe the happiest of families or the most communal of churches but it can hardly be applied to all cases, and it tends to depict organisations only in rosy hues. In the postwar era especially, such hues were those that influential elites in Japan sought to project (Mouer and Sugimoto, 1986).

The model of the Japanese organisation as a *gemeinschaft* also has some similarity with McGregor's (1985; originally 1960) Theory Y. McGregor's book was widely read in Japan and had a strong influence on Japanese personnel management. The *gemeinschaft* reasoning suggests that, because the company respects the welfare of its employees and treats them equally within the lifetime employment system, employees will willingly devote themselves to the organisation. While reality and the expectation are not in total discord, the reality has not wholly lived up to the expectation.

There are obvious differences between the Japanese and the US system, but it is the latter that has defined normal practice for so many management scientists. Most notably, in the Japanese system management salary differentials are far less, employees in the internal labour market are m the adverse rhythms of the employment cycle, and there is volvement of the rank-and-file in decision-making.

there have appeared several American studies on the relationship between personnel management and corporate performance. These studies highlight a number of features of successful personnel management, including employment security, an internal labour market, tightly defined job descriptions, participation in decision-making, extensive formal training, results-oriented appraisal and the linking of pay to performance (Schuler and Jackson, 1987; Pfeffer, 1998; Delery and Doty, 1996; Youndt *et al.*, 1996). Some of these features bear a great similarity to aspects of the Japanese model (Table 10.1), the key features of which will be examined in the sections that follow.

Table 10.1 Features of Japanese and American human resource management

	Japanese model	*Traditional US model*
Basic principle	*Gemeinschaft*: long time horizon, egalitarianism	*Gesellschaft*: short time horizon
Nature of employment and recruitment	Lifetime employment,* recruitment of new graduates	Hiring and firing or lay-off, new recruits from outside
Role definition and promotion ladder	Flexible roles; non-specialist career paths, dual promotion ladder*	Narrow-skill jobs; specialised career path, Promotion up job ladder
Placement and promotion	Internal labour market, planned rotation; promotion by ability and tenure	External market; seniority for union members, movement within in the same occupation
Appraisal and training	Frequent appraisals, abilities and attitudes measured; intensive training*	Less frequent appraisal, performance measured; skills bought from outside
Reward and competition	Pay for skill,* small differentials and frequent increases*	Pay according to job, differentiation by job
Labour–management relations	Company-wide union, cooperative	Industry union confrontational
Leadership and commitment	Participative leadership, self-control, mutual trust, commitment	Job-centred leadership, external control, low trust, compliance

* Key features.

The strong emphasis on personnel is reflected in the formal creeds of Japanese corporate organisations. For instance Matsushita's creed declares that 'Matsushita creates respectable personnel before the production of goods'. Another example is 'The company is a school to train personnel'.

The first major postwar study of Japanese personnel management by an overseas researcher was Abegglen's (1958) study of the lifetime employment system. Later, despite being only indirectly about Japan, Ouchi's (1981) *Theory Z* became one of the best selling books on 'Japanese management'. In it were described some of the features listed in Table 10.1. Earlier studies by Benedict (1946), Nakane (1967), Doi (1971) and Hazama (1976) emphasised the organisation-orientedness of Japanese employees, fostered by lifetime employment and the internal labour market system.

While some of these features are based on traditional Japanese social values, most of them did not appear until after the Second World War. For example employment was not stable before the war in the textile and mining industries. Stable employment only emerged as a result of union demands and the new needs of industry after the war. As industry began to produce technology-intensive products, skilled workers had to be trained and it became economically necessary to retain them for years. So both unions and employers were happy with a system that encouraged long-term employment and commitment.

LIFETIME EMPLOYMENT

The recruitment of graduates of high schools and universities is usually conducted in April. Companies seldom recruit by advertising at other times of the year. Once a person enters an organisation he devotes himself to it and stays until he retires at 60. (Lifetime employment is very largely gender-specific, hence the use of the masculine.) He will not move from organisation to organisation. The organisation will take care of him throughout his working life and will not lightly terminate his employment. For an employee, resignation or discharge can seriously damage his career because it calls his character into question.

There are several misconceptions about lifetime employment. First, it is not a contract, rather it is a particular way of thinking by both employer and employee. When profits decline a company will take many measures to reduce its costs, including the curtailing of dividends and the bonuses of top managers, but it will retain its employees for as long as possible. This is in contrast to US corporations, which tend to lay off employees to keep the dividend rate high, and increase the value of stock options for top managers even when laying off employees.

The lifetime employment system does not mean, however, that the staffing level is never reduced. After the oil crisis and since the recession of the 1990s many companies have reduced their workforce, but this has been done on a phased basis rather than through sudden redundancies or lay-offs. There are various ways of coping with a lower demand for man-hours: cutting overtime; transferring workers from departments with surplus labour to other departments or subsidiaries; suspending new recruitment; offering voluntary early retirement; and introducing temporary paid leave. Termination is used only as a last resort, and older people are asked to go first.

Second, as mentioned above the lifetime employment system does not stem from traditional social values. Before the Second World War, companies had little respect for their workforce. Young female workers and unskilled workers from farming families were seen only as interchangeable parts and labour turnover was high. However, heavy industries such as shipbuilding, railways, chemicals and mechinery needed skilled workers and it was important to hold on to them. It was these industries that developed the lifetime employment system (Yoshino, 1968; Nakagawa, 1977).

After the war the structure of industry changed: technology-intensive heavy industries gained importance over light industries such as textiles. Moreover the unions were reorganised (they had been banned during the war) and demanded stable employment. Democratic ideas became prevalent, in part under US tutelage, and these encouraged respect for people not only in the civil sense but also for those working in organisations. The reason for this was that many of the prewar *zaibatsu* had been seen as encouraging the war as an outlet for their production.

There are a number of advantages to the lifetime employment system: it encourages a strong commitment to the company and the job; it provides the possibility of intensive training and planned career development; it promotes favourable labour relations; it encourages the accumulation of knowledge and organisational learning; and it facilitates job rotation and flexible job assignment as well as fostering multiple skills.

However, there are also disadvantages. First, as the average age of employees increases, labour costs increase. Under the pay for the job system, age and wage are not connected, but under lifetime employment wages are governed by length of service. Second, there is a lack of fit between job and ability in the case of older employees. Employees tend to be promoted to higher-skill jobs because of length of service rather than ability. This problem can be averted by appropriate job assignment and training, that is, by separating job assignment from length of service. Third, it is hard to cope with fluctuations in demand and production or take advantage of technological changes under the lifetime employment system. Companies that can expand or cut back their workforce according to demand are more flexible. This is the usual practice in the US. Discharging those with redundant skills and employing people with newer, more useful skills reinforces the use of new technology rather than hinders its adoption. Likewise divesting unprofitable businesses and acquiring more promising ones improves flexibility.

Although organisational inflexibility is an inherent problem with lifetime employment we should not lose sight of the benefits. These benefits have even been recognised by firms that are more used to the American way of doing things. For instance Hewlett Packard, Motorola and 3M have seriously considered the benefits that lifetime employment and respect for people offer to company growth and increased sales and profits.

Organisations can cope with changes in demand and technology by changing their product mix in related technologies and by training employees. It is the accumulation of knowledge through long-term employment and job security that leads to the 'learning organisation' and makes Japanese businesses so good at incremental innovation. What happens to innovation when long-term employment and job security cannot be guaranteed can be seen in the *Sunday Times* newspaper in Britain. In the early 1980s the proposed modernisation of production meant that many members of craft unions would lose their jobs (based on the old 'hot-lead' technology) and as a result they went on a strike that became a lockout. The company could not use its newly installed equipment for many months because of the strike. In Japanese corporations such opposition rarely happens; rather workers welcome modernisation because employment is assured. For instance Nippon Steel constructed a new plant in Kimitsu near Tokyo and closed its plant in Kyushu. Thousand of employees moved the 1000 kilometres from the old plant to the new one, where the company provided the housing. Thus job security makes it easier to carry out technological innovations.

The lifetime employment system reduces the possibility of employees changing employers because doing so would marks them out as a non-company men – as inherently untrustworthy. Therefore if an employee is assigned a job that does not suit his abilities and interests he can only dig in and wait for another opportunity.

The head of *Keidanren* (the Federation of Industries) stated in an official speech in 1999 that members of top management who cannot protect the jobs of employees should discharge themselves first. Such words would sound surprising if uttered in Australia, the US or Britain, where one is more likely to see the reverse, and even if departure becomes necessary for executives because they have failed in their duties they frequently make their exit attached to a 'golden parachute' – substantial termination packages that reward them amply even though they have clearly failed their employees and the shareholders!

FLEXIBLE ROLE DEFINITION, THE DUAL PROMOTION LADDER AND THE INTERNAL LABOUR MARKET

Role definition

In Japan, role definition is typically ambiguous and the scope of jobs is wide. At Matsushita and Toyota there are no job titles and employees are assigned to a certain group that is responsible for a particular array of jobs. Who in the group does what is not prescribed. Employees carry out numerous jobs and are rotated from department to department. Even within the same department, employees are moved from job to job in order to avoid monotonous repetition. They not only operate machines but also clean and maintain them. They do not wait for someone from the maintenance department to fix malfunctions because they know how to do it themselves. Such a system is made possible by the fact that wages are not determined by the job but by status grade.

There are several advantages to such a system. First, because of frequent technological innovations the nature of jobs changes from time to time and the system provides the necessary flexibility. Second, the quality of products is improved because employees are expected to monitor and correct quality problems themselves. Third, higher job satisfaction tends to be achieved when people operate under conditions of job enlargement and enrichment.

There are problems with such job flexibility, however. First, generalist skills do not map well onto situations where highly specialist skills are required, for example in the marketing and servicing of high-tech products, production by computerised machines, assembling made-to-order products and designing new products. Hence there is a new trend to provide courses in specialist skills. In order to reward innovation and advanced skills, a larger differentiation of rewards has become necessary, introducing problems that will be studied later in this chapter.

The dual promotion ladder

In many companies there are two promotion ladders: the hierarchy of job grades and the hierarchy of status grades. As a result there are many opportunities for promotion, but wages are usually determined by status grade rather than job grade. Employees are promoted up the status ladder in accordance with length of service and merit. A minimum number of years have to be served in the same grade, after which the employee is eligible to be promoted to the next grade. However promotion is by no means automatic. The nature of the job (and the job grade) and the employee's ability and effort are all taken into consideration. This means that to some extent there is a matching of job grade and status grade, but an employee can be promoted in terms of status grades while remaining in the same job. Usually, only the status grade is employed for lower-level workers and both grades for higher or management levels.

This system is quite different from that in the US, where if an employee's job does not change then his or her grade (and wage) does not change. We can see the system in use by looking at the case of Sony, where there are three ladders. At the lower levels there is only the status ladder, with eight grades; for positions higher than section manager there are three ladders with eight grades. This system applies to all departments. Salary is decided only by status. For promotion to occur there has to be at least one recommendation by a supervisor, a test or an interview. All employees are eligible for promotion up to the ninth grade (assistant to section manager). Until the age of 29 there is no differentiation in speed of promotion but after 30 years of age there is competition for promotion. Employees have to serve a minimum number of years in a grade before they can be considered for promotion. In this way overly rapid promotion is prevented. From the lowest grade up to that of group leader the minimum stay is about two years; after that three years is the minimum tenure. Above the level of department manager there is no minimum stay period.

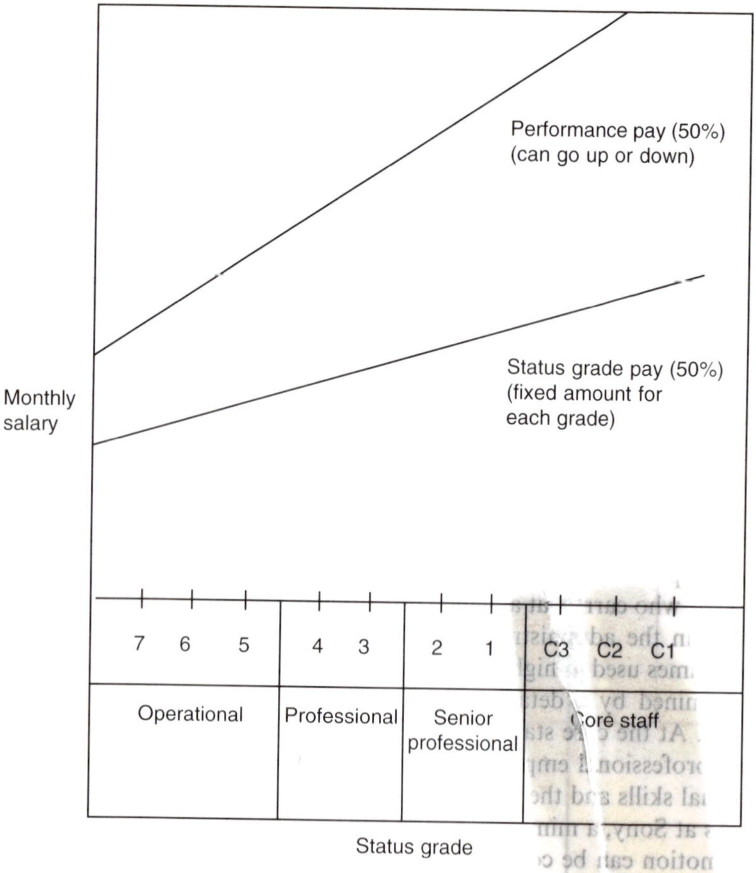

Notes: In addition to monthly pay, bonuses are paid in July and December. This amounts to about five or six months' pay. Under this system a large differential can appear.

Figure 10.2 Pay and status grade system at Toyota

There is no rapid promotion – employees are promoted by small increments, based on the gradual accumulation of merit. Thus they have to look to the long term.

Certain other drawbacks have to be balanced against the advantages. To evaluate each individual employee is more troublesome than to evaluate jobs. In addition, such evaluations can be subjective, although the use of frequent and multiple evaluations guards against this to some extent. For example, separate evaluations of one employee may be conducted by three supervisors at three successive levels, and these are then compared. At Toyota, each department head is required to evaluate 10 members of other departments. In addition subordinates are required to evaluate their supervisors by means of anonymous questionnaires.

The internal labour market

In the internal labour market system the company recruits fresh graduates and trains them for potentially any kind of job. When the skills required change as a result of changes in product mix, employees are retrained for the new jobs. This system is an important feature of lifetime employment and contrasts markedly with the 'hire and fire' system, where new recruits are hired for a particular job. If the job goes, so does the employee – thus restructuring is easy. Job titles and the associated duties are clearly specified and employees specialised in certain skills, so if they move they move to similar jobs. Wages are determined by the job or the job grade. If there is a seniority system, young employees with a short period of service are laid off first, perhaps to be hired again when needed. They lose seniority when they move to another company. When there is a vacancy in a higher position the person with the greatest seniority (the longest service) will be promoted automatically (although some jobs, for example train driver, also require appropriate qualifications).

The internal labour market makes it possible to retain employees and it creates high morale. Broader and deeper knowledge is accumulated during the course of lifetime employment, and internal growth is facilitated. While these are undoubted advantages there are some drawbacks. For example drastic changes in product-market strategy are not possible because there can be no quick reshuffling of capabilities, which is one reason why there is such an emphasis on long-term planning. However related diversification is possible.

APPRAISAL AND TRAINING: THE LEARNING ORGANISATION

Japanese corporations put great emphasis on enhancing the knowledge of their employees, because under lifetime employment it pays off, organisationally, to do so. It also serves to enhance respect for people. As stated earlier, part of Matsushita's creed is 'Matsushita puts respect for personnel before the production of goods'.

Table 10.2 Example of an appraisal schedule

(1) Name and department
(2) This year's role
- Expected role
- Detailed duties
- Important subjects, goals and schedules
- Required knowledge and capability

(3) Evaluation of performance and progress
- Self-evaluation (October)
- Supervisor evaluation (October)
- Self-evaluation (April)
- Supervisor evaluation (April)

(4) Capability appraisal (February)
- Technical knowledge
- Problem-recognition and planning ability
- Implementation power, mobilisation ability
- Support of others and leadership power
- Other

Employee appraisal: the case of Toyota

Toyota appraises all its employees three times a year: once for promotion and wage increases, and twice for bonus payments. There are detailed evaluation standards. In the case of technical skills, the required skills are specified for each career path (there are about ten career paths). For conceptual and human skills there are specifications for each rank. An example of an appraisal schedule is provided in Table 10.2.

The process is similar to management by objectives (MBO). The employees duties, goals and required skills are defined in April, taking into consideration the department's goals. Performance evaluation against the set standards is conducted in October (and used for the winter bonus) and again in April (when it is used for the summer bonus). The process involves both the supervisor and the person being supervised, as does the capability appraisal, which is conducted in February. The grades A, B, C, D and E are used for marking. The direct supervisor makes the first appraisal, using two-way communication: the subordinate explains his performance and then the supervisor explains the evaluation, and informs the employee of the future efforts and capabilities needed. The department head conducts the secondary appraisal and the head of another department does the third, then the subordinate completes the appraisal by completing a questionnaire.

Table 10.3 Evaluation items and their weighting at managerial and core staff level (per cent)

	Weight		
	Manager	*Core professional*	
Specialised knowledge and skill	0	50	
Issue finding and new task creation	20	10	Conceptual skills
New task implementation	30	15	
Management of groups	20	10	human skills
Encouragement of subordinates	20	10	
Respect from others	10	5	
Total	100	100	

At the managerial and core professional levels the appraisal items are weighted (Table 10.3). For managers conceptual skill is given a very high weight (about 50 per cent). For core professionals, technical skill is highly weighted (50 per cent), while conceptual and human skills are weighted at 25 per cent each. Comparing the results with the goals determines the performance evaluation. Self-appraisal is done first, then the supervisor makes an evaluation, and then two-way communication is carried out. The technique known as 360-degree appraisal is also practiced. Managers of other departments and the subordinates of the person being appraised also participate in the appraisal.

For promotion, in addition to the regular appraisals the recommendation of department heads is needed (and competence in the English language for higher professionals). There is a limit to the number of higher-rank positions, but slow movers in the promotion stakes are nonetheless encouraged to make the effort to catch up.

Training and learning

Off-the-job training

Knowledge and attitudes are derived from others through experience and inculcated patterns of thinking. There are three ways in which employees acquire specific knowledge and develop the desired attitudes. First, orientation training is provided to the new recruits who join the company in April. About six months are spent on this, beginning with classroom training and then on-the-job training in marketing and production. During this time not only technical knowledge but also the company philosophy is communicated and inculcated. Second, training occurs before promotion. At Toyota training is also provided after promotion to each of the status grades (Figure 10.2). The personnel department mostly provides training in conceptual, organisational and interpersonal skills, while the other departments provide training in technical skills. Third, there is occupational training in finance, procurement, marketing, production, design, maintenance and customer service. There are few statistics on the duration of such training, which is undertaken outside the company, but it is estimated that is at least three days a year per employee.

Training in decision-making

Honda provides extensive training in quality control, teaching the PDCA process (plan, do, check, action) and 'fish-bone' analysis using histograms and control charts. Honda also provides a course on the Kepner–Tregore method (Kepner and Tregore, 1965). These activities are useful for improving decision-making.

Following the principles of quality control and the Kepner–Tregore method, Honda proposes three steps in arriving at a decision: goal setting, information collection and idea generation. First, employees set the future goals of their jobs. Second, they analyse their present goals using the 'three Gs' – information collected at the operating front (*genchi*), at the actual situation (*genbutsu*) and in terms of the issues at hand (*genba*). Employees are required to forecast what will happen if no action is taken and the likely actions of the company's competitors, and then what should be done in the future given the situation just analysed. Such training in decision-making also improves learning ability and action taking.

Learning from subordinates

In order to minimise social and power differentiation in the organisation all Honda employees – from plant manager to shop-floor worker, office manager and clerk – wear white uniform suits. This practice applies not only in Japan but also at Honda's overseas facilities. The president and other executives visit all the front lines at least once a year to obtain information from production staff and sales employees.

This information is used to formulate new ideas, develop plans and implement them throughout the company.

Self-training

Companies support the expense of seminar attendance and correspondence courses. Promotion is premised on success by candidates in competency tests which stimulates their learning. Some companies offer long-service 'refreshing vacations' to their employees, for example Kirin Brewery awards a one-month paid break to employees with 20 years of service so that they can refresh their knowledge – a similar notion to that of an academic sabbatical.

Training through job rotation

New recruits are rotated among at least three departments – for example the sales office, the production plant and the accounts department – for six months or more. Even science graduates recruited to work as researchers in laboratories are required to spend time in the sales and production departments. This rotation enables employees to acquire broader knowledge and establish relations with different departments. Employees of several years' standing are also rotated to different departments to broader their capapabilities. As wages are determined by status rather than job, rotation is considered acceptable, even at the operating front. For promotion to chief of section, experience in more than three functional departments is required. Recently however, rapid technological change means that specialists (rather than generalists) are increasingly required, so the system is changing (this will be discussed in a later section).

Learning from failure

When an experiment fails, people tend to hide the fact because it could be seen as reflecting badly on their ability. Yet it is important to learn from failure and thus avoid similar failures in future. At Sony the development department studies failures as well as successes to discovers the dynamics and causes of each. Failure is not punished – freedom to fail is an important factor in the encouragement of experimentation.

In 1958 Canon developed a new machine called the Syncro-reader, a disk that could be used for recording at home. The company constructed a large factory of 250 000 square metres in the suburbs of Tokyo. However, in spite of the praise accorded to the product by the newspapers it did not sell because it was too expensive. The president of Canon put the blame on himself, saying 'I was foolish', a statement that had a great impact on the people at Canon as it indicated that failure would not be punished as long as it was the result of a real effort to succeed.

Learning through thinking

The suggestion system. Our survey in 1995 showed that about 80 per cent of Japanese corporations employ a suggestion system and that between 3.5 and 24 ideas per person per year are presented. New ideas by employees are written on suggestion sheets and evaluated by a committee. Monetary rewards range from 1000 yen to 160 000 yen, depending on the quality of suggestion and the policy of the company. In many companies the names of those who have presented many good ideas are displayed in a 'roll of honour' posted on the wall. Suggestion systems encourage employees to change their traditional way of thinking about their jobs and thus to present new ideas.

We shall briefly discuss two examples where new knowledge was produced through the suggestion system. At a Suzuki production line for cars an idea was presented for ensuring models did not get shipped with loose bolts. Chalk was used to mark the hole for the bolt fastener, so fastened bolts had a coloured mark while unfastened bolts did not; thus loose bolts could be avoided through simple visual inspection. The second case comes from the TV assembly line at Matsushita. A worker found that defective products appeared in some cycles and she suggested that the vibration caused by trains running on a line near the factory may have been the cause. Indeed she was correct, and the factory was able to take appropriate measures to reduce the defect rate.

Quality circles. Quality circles are very popular among Japanese corporations – about 80 per cent of companies have quality circles compared with about 40 per cent in the US and Germany in the 1980s (Beaumont, 1993). A quality circle is a group of employees who meet regularly to solve problems that arise their work area. Generally, six to 12 volunteers from the same work area make up the circle. The members receive training in problem-solving, statistical quality control and group processes. Quality circles generally recommend solutions for quality and productivity problems, which management may then implement. A facilitator, usually a specially trained member of management, helps to train the circle members and ensures that things run smoothly. Typical objectives of quality control programmess include quality improvement, productivity enhancement and employee involvement. Circles generally meet for four to eight hours a month in company time. Members may be awarded recognition but rarely receive financial rewards (Lawler and Mohrman, 1985).

The transmission of ideas: morning meetings

The information provided at the morning meeting might include the plan for the day, information on the present situation, or issues facing the broader company and plant. The passing on of this information is intended to stimulate or generate new ideas. At the morning meetings at Matsushita plants, after a speech by a supervisor an employee is asked to make a short presentation. The subject matter is left up to the speaker. This format is followed every morning at all locations, even at subsidiaries in foreign countries. It stimulates new ways of thinking.

Providing information on the past performance of the company and the plant is effective because this feedback functions as a confirmation of achievements. As stated in the subsection on learning from subordinates, the metaphorical 'low walls' between departments and individuals facilitate idea transmission. At production plants, everybody wears the same uniform, there is no differentiation between blue-collar workers and white-collar workers and the plant manager frequently visits the production line. Office staff and the head of department work together in one large room, so power differentiation is minimised. Cooperation between departments is facilitated by job rotation and lifetime employment, which also improve reciprocal relations.

REMUNERATION

As already mentioned, remuneration is mostly determined by status grade. A typical example of a salary schedule is shown in Figure 10.3. There are different wage progression lines for each status grade. Higher grades have a higher line. Wages are increased every year. In many cases there is no maximum rate, and wages reflect not only grade but also individual merit. There are three occasions for wage changes: promotion up the status grade, the annual wage increase and overall wage increases.

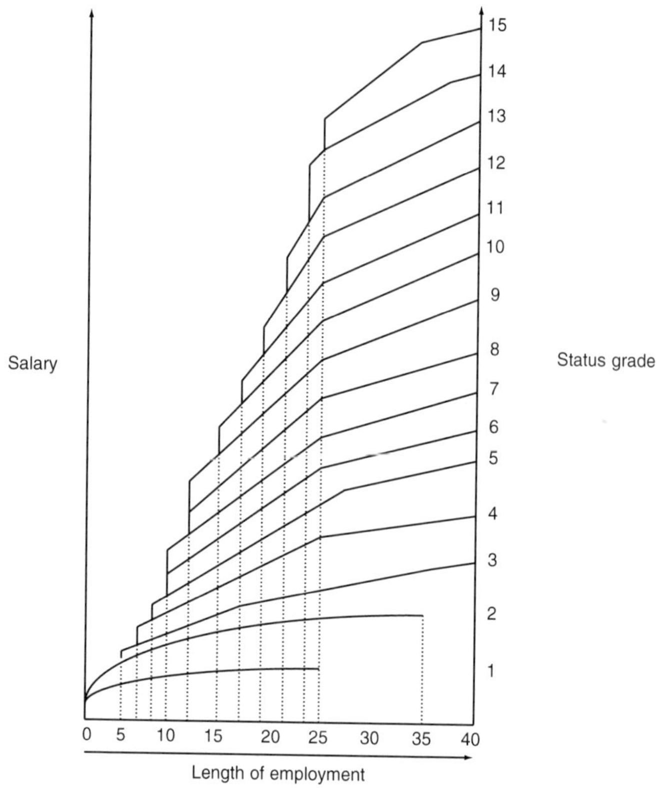

Figure 10.3 Example of a salary schedule

The method of payment is the same at every level of the employee hierarchy and wages are paid monthly. A large bonus is paid twice a year – about two months' wages in the summer and about three months' wages in December. The bonus is related to company profits and personal performance. Before the Second World War there was a large difference between the bonus paid to managers and that paid to rank and file workers, who received only a very small bonus. At that time the wage system was largely based on the job and the job grade.

What are the merits of the present remuneration system? First, there are frequent wage increases but only a small differentiation between levels, which provides a strong incentive to work hard. Second, the wages paid to new and young workers are low but they increase with length of service as their family responsibilities and cost of living increase, so there is a balance between performance and reward in the long term. These low wages ensure that young people will not be discriminated against by prospective employers who do not want to pay an adult wage to inexperienced people who lack the skills of more senior workers. Third, it encourages lifetime employment, because the longer the tenure the higher the wage.

However, there are some problems associated with the system. For instance labour costs increase as the average age of the employees in an organisation increases. Also, the strict seniority rules mean that young employees may receive a lower reward than their contribution merits, for example those with special expertise to contribute – perhaps in new areas such as e-commerce.

A typology of promotion and wage systems

A representation of the promotion and wage system is provided in Figure 10.4. The merit system emphasises capability and provides for the rapid promotion of those most able; the corollary is that it also allows for the sacking or demotion of the less able. From the viewpoint of the employee, the system does not offer stability but does reward merit. Opposed to this is a system based on nepotism or entrenched social groups. In such a system capability is ignored and only those from, say, a certain family, social class or caste will be promoted. Ability will not be taken into consideration.

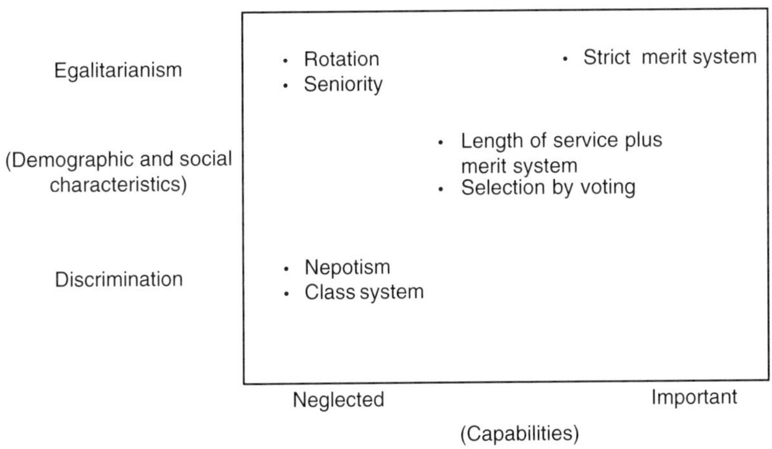

Figure 10.4 Models of promotion and wage determination

The 'length of service and merit' system fits between these two extremes, the Japanese version of which is a blend of egalitarianism and reward for merit. It is fair and also provides stability for employees. The US personnel system is closer to the merit system.

CONTEMPORARY TRENDS

New problem areas

Changes in the economic and business environment and the predisposition of employees have created various new problems for organisations and required them to adopt new practices, as shown in Figure 10.5. First, the economic downturn has increased the financial burden imposed by the lifetime employment system. Second, mega-competition, globalisation and changing consumer demand are necessitating the use of specialised rather than generalist skills. Third, the fall in the birth rate has resulted in a rise in the average age of employees, thus increasing labour costs. Consequently some features of Japanese personnel management, such as lifetime employment and the emphasis on general skills and egalitarianism, have to some extent been modified.

The new practices that have emerged are flexibility in the employment system, increased emphasis on specialised skills, greater respect of individuality and enhanced creativity and learning.

Terms of employment

In companies where lifetime employment continues to be basic policy, if an employee wishes to take early retirement the company will, for instance, pay an increased retirement allowance of more than one month's pay per year of tenure. Sometimes companies help employees move to other companies by training them appropriately. (It is important to remember that Japan does not have a well-developed welfare state with an adequate retirement pension. Essentially, individuals have to fund their own retirement.)

Matsushita

Matsushita has adopted the following plan. Firstly, it advises all employees to attain specialised skills so that if necessary they can obtain employment elsewhere. At the age of about 50 years employees

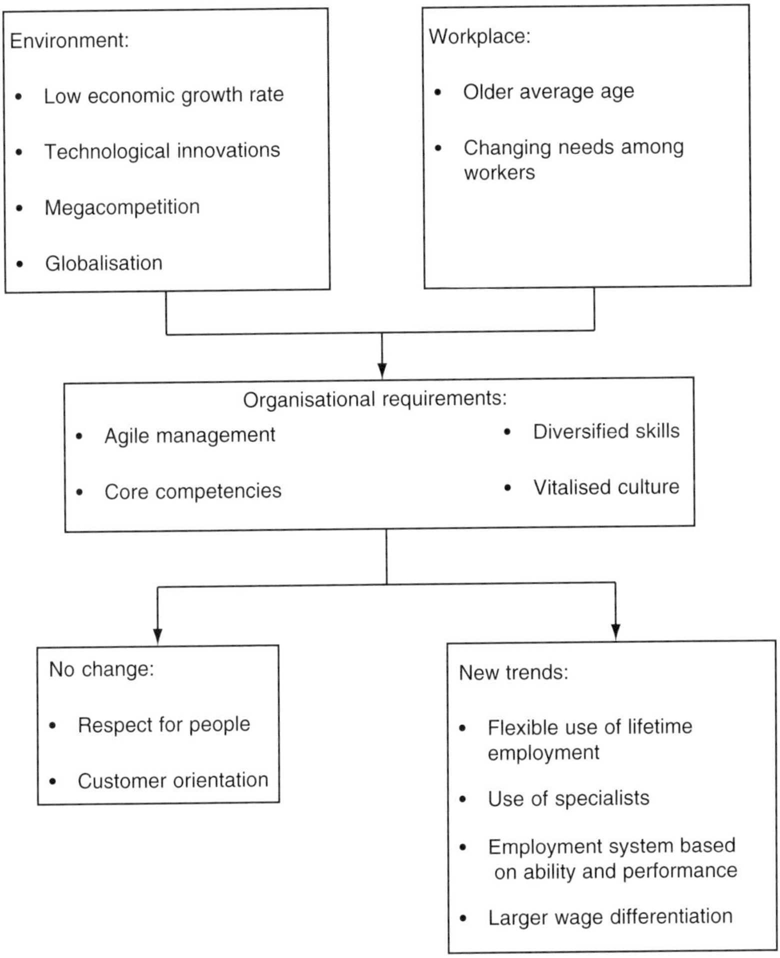

Figure 10.5 Factors generating change and their consequences

become eligible for the 'career development holiday scheme', whereby those who are willing to take part are introduced to another company. An increased retirement allowance is also offered to encourage early retirement. New career seminars are held to assist employees to make an inventory of their capabilities and competencies and to help them to develop additional skills. For those who stay on, when they are about 58 years of age the company asks them whether they wish to continue working for the company or to retire. At the age of 60 all employees are encouraged to retire, but if they continue to work they are given new jobs. Alternatively the company may help them to find work with an affiliated or unaffiliated company. The new jobs tend to be simpler and lack the accrued benefits of the seniority system. Those who continue to work are paid at the market rate, but they must retire at 65.

Tokyo Gas

Tokyo Gas supplies 8 450 000 customers, employs 12 000 people and has about 3000 contract workers. The company has introduced six 'second life paths' as part of its formal employment plan. First, under the dispatched employment path, employees who are sent to affiliated companies before 60 are employed by the dispatched company at the age of 60 on one-year contracts until they are 65. Second, under the work-sharing path, 60 year-old employees work three days a week as contract employees for about one third of their previous salary. Third, under the free-contract path, retired employees with special skills, such as accounting, electrical equipment maintenance or information systems expertise, work as contract employees when the company requires them. Fourth, employees aged 55 or over work four days a week while preparing for a second working life – they are paid four fifths of their previous salary. These employees then try to find another job through their own efforts. Fifth, employees aged 50 or over can try to find a new job, but the company offers to pay an increased early retirement allowance or the same pension they would have received had they been 60. Finally, there is the standard path, where employees retire at 60 and receive the standard retirement allowance or pension.

In order to support its employees' working lives the company provides the following services. First, it provides regular 'regeneration breaks'. At the age of 30, 40 and 50 employees are able to take long holidays using their accumulated holiday entitlement. At the age of 50, special holidays leave and a 300 000 yen travel token is provided. Second, at the age of 45 employees attend a seminar aimed at helping them to further their careers. Third, at the age of 50 employees have a 'career development interview' with their senior manager about their opportunities for career development within or outside the company. To this end the company provides a 'Career 50 guidance book', which explains options for the second life path. Fourth, another type of seminar is provided to employees who have reached the age of 53. At this two-day seminar, which is also attended by the employees' wives, the participants receive tutorials on the various retirement options, talk with each other in small groups, and listen to presentations from successful retirees. Fifth, the 'new life seminar' is provided for employees aged 59 – this is similar in nature to the seminar just described. Finally, the company provides 'second life support', whereby personnel staff provide employees with consultation services.

These examples demonstrate how companies are now trying to provide various forms of support so that employees can find a meaningful second working life. The retirement counselling system is also one of the features of lifetime employment. It serves to underwrite a paternalism that regards employees, up to a point, as members of the corporate family. As we have seen, that family is not all-embracing when it comes to its older members: we shall see next that it is not necessarily for other employees either.

Temporary employees

Widespread use is made of temporary employees in Japanese corporations. These temporary employees not only secure the employment of permanent employees but also provide the company with necessary skills, sometimes high-level skills, thus enabling the company to concentrate its resources on building up its core competence. Nationwide the proportion of temporary employees is about 20 per cent (8 per cent of male workers and 36 per cent of female workers in 1995), but the numbers and composition vary from company to company.

Temporary employees can be classified into (1) temporary outsourced employees, (2) part-timers (about 0.6 million in 1995), (3) contract employees on contracts of between one and three years, (4) short-term employees and (5) employees of other companies (affiliated companies, cooperating companies or companies further up the value chain). The latter are engaged in simple jobs such as assembling, cleaning or general maintenance. Some temporary employees may belong to the company union.

Recently out-sourcing has been on the increase, mostly in connection with computer operations, accounting and clerical work for personnel management. Out-sourced employees are used to carry out expert jobs that are not core jobs for the company. The employment status of these non-regular employees is not necessarily shaky as they may be permanently employed by other companies or be taking a short break from farming. Temporary employment should not necessarily be considered an unfair labour practice because it conforms to the particular needs of some workers, for example some people do not want to be employed on a permanent basis and some do not like to work for eight hours a day.

Specialised career paths

Japanese companies have tended to emphasise the employment of generalists because the period of employment is very long and wages are not determined by the job. Given the lengthy commitment to core workers, a system that employs generalists is more flexible. However the technological level of most jobs has risen and companies now need more specialist skills. In addition, employees are more inclined to aim for self-realisation through higher level jobs.

The career paths available at Toyota are shown in Table 10.4. New recruits receive general training for one year and then apply for a career path. The company considers the application and decides the appropriate path, looking in particular at the applicant's academic background and the degree of need there is in each specialised field of the company. The personnel department usually takes such decisions. Within these career paths, employees can advance up the status grade system if entitled, as far as head of department.

Changing a specialised career path is possible at Toyota and is applied for on a 'self-statement sheet'. There are two kinds of transfer between career paths: permanent transfer and rotation for training –

Table 10.4 Specialist career paths at Toyota

1. General management:
   corporate communications, personnel, finance and accounting, information, procurement
2. Marketing:
   domestic sales, after-sales service, overseas management, overseas sales, overseas services
3. Technical development:
   product planning, design, bodywork, engines, chassis, experiments
4. Production engineering:
   machine operation, bodywork, assembly, casting etc.
5. Production control:
   production control, production planning, supply chain management
6. New business development:
   ITS, new business development, information management, shipping housing

with the latter the employee eventually returns to his original career path. Promotion within a career path is carried out by the appraisal system, as stated earlier. At the higher level, employees are divided into the administrative path and the special staff path. Wages are determined by the grade in the status ladder of each career path, with no differentiation of wages between paths.

Tokyo Gas (12 000 employees) has 15 specialised career paths that are grouped into five sets: (1) sales engineering, (2) marketing, (3) production and maintenance, (4) basic technology and (5) management support. Allocation to these paths is determined by the subject the employee studied at university or high school. After three years of observation, they are interviewed in the fourth year and allocated to one of the sets. There are opportunities to change career path after 7 and 12 years of employment, and at the age of 50. In their seventh and twelfth years employees receive training in the company's career development programme. Each course has training and support staff.

New posts

New posts are advertised throughout the company and applications are sent directly to the personnel department, without the need for permission from the section head. Examples of new posts are jobs in foreign countries, other places in Japan or in new businesses (for example Sony set up a new life insurance business). About 2500 of Sony's employees have moved since the introduction of this system in 1966. At Matsushita about 100 people a year move as a result of the system. The system differs from applications for promotion to a higher grade in the status ladder or a higher post. It is initiated on the demand side and mostly involves horizontal moves. Other companies are adopting a similar system, the advantage of which is that employees can choose their own opportunities; thus, potentially, greater motivation and respect for individuality can be achieved. If a superior does not take good care of his subordinates, it is sometimes the case that they will rapidly transfer out of that section.

Conceptual skills and an innovative attitude

The appraisal system is changing to place more emphasis on conceptual skills and an innovative attitude. In the past, appraisal was based on current performance and a cooperative attitude towards colleagues.

(see Table 10.3 and the accompanying discussion) Evaluation is now based on a goal statement, which includes expected roles, important tasks, related corporate policies and goals.

At Kirin Brewery (Table 10.5), strategy formation has a high weight in the evaluation of higher level managers. Evaluation is based on goal statements, a kind of MBO system. The items in the goal statement include general goals, methods of achieving those goals, schedules, degree of difficulty, self-development plans, desirable future jobs and self-assessment of basic skills. These evaluation schemes emphasise conceptual skills and highly specialised knowledge, while allowing employees to express their individuality. They serve to strengthen core competencies in terms of organisational requirements, and to foster a learning organisation.

Feedback on the performance appraisal

About 30 per cent of large corporations (Ministry of Labour, 1997) have started to inform employees of the results of their performance appraisals. Toyota does this only with higher-level managers and senior specialist staff, not lower-level employees. Matsushita transmits

Table 10.5 Appraisal system at Kirin Brewery (appraisal items and weightings)

		Weight		
	Level of achievement	*Management level 1–2 (director of department)*	*Management level 3–4 (assistant director)*	*Management level 5–6 (manager)*
Strategy formation and innovation	Remarkable	100	50	20
	Good	50	30	10
	Moderate	0	0	0
Achievement at present job	Sliding scale, full marks when remarkable	0–50	0–50	0–50
Self-development or development of subordinates	Remarkable	50	50	30
	Good	30	30	20
	Moderate	0	0	0
Total		0–200	0–150	0–100

Note: MBO is used for evaluation.

the results to most employees above section manager, either in writing or orally.

Appraisals have tended to be rather subjective, and in order to make them more objective a number of additional people have been drafted in to the evaluation process. At Toyota, departmental managers appraise about 10 people from other departments and subordinates evaluate their superiors by means of secret questionnaires. The evaluation compares the same grades in the status ladder. The merit of this is that it ensures fairness and minimises bias (on this see Beaumont, 1993).

Increased differentiation in rewards

Japanese personnel management used to emphasise egalitarianism, rewarding long service and giving frequent wage increases and promotion in small increments (there was very little differentiation between levels). This system provided a strong incentive to work hard when the economy was growing strongly and the opportunities for promotion and wage increases were abundant. But when the economy and companies' sales declined the defects of the system became evident, increased differentiation appeared necessary and enhanced specialist knowledge became more highly valued.

At Toyota, wages were composed of basic pay, reflecting capability (about 40 per cent), age pay (about 20 per cent) and status grade pay (about 40 per cent): the system was changed to that shown earlier in Figure 10.2. Status grade pay is a fixed amount for each grade. It accounts for about 50 per cent of remuneration and reflects capability, as determined by the assessment items listed in Table 10.2. Performance pay is determined by four evaluation grades and its increase or decrease depends on the employee's performance appraisal for the preceding year, not his accumulated appraisal results. A bonus (about 5 months' pay) is also paid (using MBO). Performance is evaluated each May and October.

In this system the pay differential between people of the same age group can be more than one million yen, which is a direct result of changing the appraisal method from comparing same-entry-year employees to comparing same-grade employees. Hence the new system represents a move away from pay based on length of service to pay based on ability and performance.

The annual salary of managers is determined once a year and calculated by means of the following formula: (monthly pay × 12

– bonus) / 12 = monthly payment. The salary differential between employees in the same grade can be more than two million yen.

At Matsushita, annual salaries are paid on a monthly basis to non-union employees and managees above the grade of section manager. The basic salary (about 60 per cent) is determined by status grade. There is a rate range, reflecting capability and performance. Performance pay (about 40 per cent) corresponds to the bonus. The amount is determined in June and is notified to employees at the same time as the appraisal results. The salary differential between employees of the same status can amount to more than two million yen a year. Salaries can also go down, depending on performance – something that never happenned in the past.

At Sony managerial salaries are composed of basic pay (about 70 per cent) and performance pay (about 30 per cent). Basic pay is determined by the grade in the status ladder and the rate range for each grade. Salaries increase every year but the amount of the increase reflects the appraisal results. Performance pay, which can rise or fall, is determined by the performance of the product division (about 15 per cent) and the performance of individuals (about 15 per cent). As at Toyota and Matsushita, the differential between employees in the same grade can be more than two million yen a year.

EMPLOYEE COMMITMENT

Japanese workers work hard and are willing, sometimes obliged, to work overtime. It is not unusual for white-collar workers to stay at work until late in the night. Furthermore when Japanese workers take a holiday they do so not when it suits them but when it suits the company (Table 10.6). To Westerners this may seem surprising, shocking even, but the story gets even worse – most Japanese workers do not fully use their holiday entitlement but accumulate it until their retirement, or until the company buys it out. They rarely if ever take false sick leave: the concept of the 'sickie' is not known. Most Japanese workers spend a lifetime with one company. In turn the company tries to protect their jobs, for example by not transferring production to lower-labour-cost countries and closing more expensive plants in the home country.

However comparative studies have revealed that Japanese workers are among those least satisfied with their present jobs. For example the two surveys reported in Table 10.7 found that Japanese employees are less satisfied with their jobs than their counterparts in other countries.

Table 10.6 Two models of employee commitment

Japanese model	*Traditional model*
Attitude towards work:	
Committed to job and organisation	Alienated
Hard work orientation	Minimum work orientation
Concern for total process and quality	Segmented responsibility
Holidays:	
Taken at the convenience of the company	Taken at employee's discretion (sometimes for up to one month)
Employees do not fully use their entitlement	Full use is made of holiday entitlement
Employment relations:	
Lifetime employment	Job hopping
Company protects jobs	Hire and fire

Likewise, in a survey conducted by the Leisure Time Development Center in 1995 only 30 per cent of the 1011 Japanese respondents said that their job was very important to them, compared with 61 per cent of the 1839 respondents in the US. However one should put a Japanese spin on these figures. They do not demonstrate that Japanese workers could not care less about their work but that there are many opportunities in Japanese organisation for promotion and wage increases, thus workers are not satisfied with their present jobs because they are always looking to the future.

Table 10.7 Survey of job satisfaction (per cent)

	Very satisfied/fairly satisfied	*Not satisfied/completely dissatisfied*
Survey 1: NHK (Japan Broadcasting Authority, general survey, 1980)		
Japanese (3600)	64.4	35.0
American (1680)	86.6	12.5
	Satisfied / fairly satisfied	*Dissatisfied*
Survey 2: Bureau of General Affairs (youth survey, 1989)		
Japanese (appr. 1000)	45	35
American (appr. 1000)	82	18

Note: Percentages do not add up to 100 because of other responses.

Criticisms and misconceptions

Japanese workers are driven to excess

It has often been argued that the Japanese system enforces hard work. From the Japanese perspective there is nothing wrong with hard work. It only becomes a problem when the balance between working life and private life is out of kilter. The reason for working hard is that people seek to take advantage of the opportunity for frequent, albeit small, steps up the promotion ladder and wage increases. From this perspective, the Japanese countercritic would say that if hard work was enforced, employees would not choose to stay with the company and life expectancy would be very short (actually it is the longest in the world). Still, the matter of the absence of a welfare state would need to be addressed by the countercritic as well – volition is not divorced from material reality.

Low trust

Many commentators suggest that there is no trust between the strongly hierarchical tiers of employees in Japanese companies, and that because of lifetime employment, employees cannot escape from authoritarian superiors.

Trust is built up over time and means that one can predict others' behaviour and expect that others will treat one well in difficult situations. Trust is a core concept in the *gemeinschaft* conception of organisation and it is a strong feature of Japanese corporations. Japanese workers perform high-quality work without detailed rules and specific orders. Orders tend to be few and rather ambiguous, but workers are still able to carry out precise work. Moreover decisions are arrived at by consensus. Japanese workers are willing to carry out work over and above their (admittedly somewhat imprecise) role descriptions, even if it is not ordered. After five at night, work groups often get together to drink beer and exchange views. Good cooperative relations exist between the company and the union and the union often resolves problems that arise with labour relations. All of this may seem surprising to Western observers, but it is the way that many Japanese organisations operate.

Lifetime employment is supported by many non-regular employees

Japanese companies often have more than 20 per cent of temporary employees. In particular, women's tenure is short and in small businesses turnover is high. Out-sourcing, part-time and contract employment are increasing. However, at the same time workers' needs have become more diverse and it could be argued that the existing system serves to harmonise the needs of both workers and the company. For example married women might not want to work an eight-hours day and would to prefer work part-time, and some programming experts may prefer to work for a consultancy firm because of the greater variety of technical problems they encounter.

Lifetime employment as a 'Catch-22'

Under the lifetime employment system, moving from one employer to another in order to match one's interests with one's job is difficult and reflects poorly on those who try to do so – it is almost a 'Catch-22' situation in that no company wants to employ someone who leaves his present employer in order to join another one.

However in large corporations there are a large variety of positions and companies try to fit the interests of the employee with the nature of the job. One way of doing this is through the 'self-statement' system, which allows employees to state which jobs they would like in the future. Also serving to increase the variety of jobs available are job posting, the establishment of specialist career paths and product diversification. That workers stay with and feel 'bonded' to one company does not mean that they are in bondage.

Personnel appraisal is both too bureaucratic and too subjective

As discussed earlier, appraising employees is more complicated and more time consuming than evaluating jobs. Because of the two promotion ladders, yearly wage increases and two bonus payments, appraisal has to be carried out two or three times a year for each person. In addition there are many evaluation items, and thus some critics suggest that appraisal is both subjective and bureaucratically unwieldy.

Certainly, the 'respect for people' policy requires there to be a complicated appraisal of personnel for a variety of rewards, but certain procedures are followed to make the appraisal as objective as possible, including written tests, the 'management by objectives' system, self-evaluation, the observations of subordinates, appraisal by more than three persons and sometimes the use of '360 degree appraisal', whereby the appraisal is fed back to those being appraised (about 30 per cent of large corporations – those with more than 1000 employees – practice the latter, according to the Ministry of Labour (1997).

SUMMARY

Japanese companies treat their employees as though they are members of a corporate community or *gemeinschaft*. The company creed at Matsushita, for instance, declares that the company 'puts respect for people before the production of goods'.

Lifetime employment is not a contract but a relationship of trust. It also enables organisations to accumulate knowledge. Having two promotion ladders increases the opportunities for advancement, thus reducing the desire to look elsewhere for work and enhancing employee morale

The scope of jobs is wide and employees are willing to take on extra activities. This allows the flexible mobilisation of manpower, as well as the rapid exploitation of new opportunities that emerge through technological innovation. There are few problems with job demarcation because workers are highly committed to the organisation and because remuneration is determined by status rather than job.

Employee training is emphasised under lifetime employment. Recently the emphasis has switched to specialised technical training.

Employee appraisals are carried out two or three times a year, and there are many opportunities for promotion and wage increases. In the past appraisals tended to be subjective. To make them more objective, new procedures have been introduced and a larger number of people are involved, which also makes the process more bureaucratic. Recently, appraisals have begun to focus more on conceptual skills.

Unions are organised on a company basis and tend to be cooperative rather than confrontational. The union sometimes recommends a product-market strategy to the company, as well as implementing many aspects of personnel policy. About 16 per cent of top management have served as union leaders.

Recent trends include the following. The number of part-time, outsourced and other non-regular employees has increased. In part, this is a result of a diversification of the needs of workers, and so cannot be seen as the collapse of the lifetime employment system. It can also be seen as a type of work sharing to prevent unemployment. Many successful companies have introduced re-employment courses for their employees, while keeping the lifetime employment system. This is one of the features of the 'respect for people' policy. Specialised career paths have been introduced to cope with technological innovations (traditionally, generalist skills have been emphasised). This has also served to strengthen the internal labour market. In order to encourage employees to improve their specialised and conceptual skills and performance, larger differentials in promotion and remuneration have been introduced. Long service has come to play a lesser role than merit in this respect.

Japanese employees work hard and for much longer hours than the OECD norm, and are highly conscious of the importance of product quality. It is a common misconception that Japanese workers are forced to work hard, that there is no trust between workers and superordinates, and that appraisal is overly subjective as a result of the internal labour market system. However some successful American corporations, such as 3M and Hewlett Packard, are now adopting a policy that combines stable employment with an internal labour market.

11 Conclusion

To the outsider there is much about Japan that is puzzling, even mystifying. Few things are quite what one would expect. For an observer used to the dictates of Western management theory, with its stress on the supremacy of competitive and market-oriented institutions, much of the way in which the Japanese do business will seem wrong. The state will seem too directive, the banks too powerful and shareholder value too poor. The internal labour market will seem a costly waste of resources and a source of far too much organisational slack. The importance placed on creeds will seem naive – even the most uncritical would argue that it is not very likely that a few words could capture the complex relationship that people have with the world in which they work.

These outsider criticisms are readily answered. Long ago, before much was known about comparative economics and management, Karl Marx's *Capital* warned nineteenth-century economists not to assume that the current political economy – nascent industrial capitalism in this case – would remain an eternal verity. Comparative analysis only underscores this point for those who are able to see it – where seeing it means being able to work from the inside out rather than the outside in. This strategy has guided our methodology in this book. Its insights and analysis are firmly based in an insider perspective on a complex institutional set of patterns that are, in the ways we have sought to elaborate in this book, distinctive and different from those that are usually assumed (and are often normative) in much traditional theory.

Although much traditional theory works from the inside out, it works from a different inside – an interior that is much more like that of the Anglo-Saxon economies, as seen in its most developed form in the United States but also evident – with local differences – in other countries of the former British empire, such as Australia, Canada and New Zealand. And while many of the studies that have been conducted in the traditions of this analysis – and we have cited a number in this book – are helpful in understanding particular institutionally authentic aspects of Japan, we believe that some things have been left untouched. And these, we contend, are capturable only by living in and observing its culture, and by deep and lengthy reflection and constant translation backwards and forwards from the specificities of the local to the generalities of the major international constructs at work in that field. Science is a matter of translation – from place to place, text to text, inscription to inscription, from field setting to data collection to journal notes, from journal notes to journal article. In our case it was also a matter of translation more literally, as Clegg, a Western observer, sought to make sense of the sense that Kono made of his country, his culture and his institutions. Many of the concerns that a critical observer would bring to analysis seemed to be lacking from the sense that is made in Japan. The harsh disciplines of capitalism – such as the vagaries of the labour market, the competition for mergers and acquisitions that awaits underperforming assets, and the ruthless hiring and firing of people that characterises extremely competitive firms in extremely competitive economies – are absent for those inside the internal labour market of large Japanese firms.

Our book focuses only on these firms and thus, by implication, the internal practices that characterise them. We have not addressed the world of work outside these firms, the myriad second-, third- and fourth-tier subcontractors, even fifth-tier subcontractors, where many of the everyday injuries of the Japanese economic system – economic, gendered and emotional – are hidden from view. Our window is on a world of economic and organisational privilege compared with life in a sweatshop or a hazardous paint factory. Hidden from view, such damage is not central to the concerns of big firms. Within these firms the practice of power takes a different hue from that in other places. Once prospective employees have managed to be selected by one of the big companies, such as those we have focused on in this book, then not only can they count themselves very fortunate, but they can also reasonably expect to be taken care of for the remainder of their career with that company. And while this expectation may be changing, it is not changing at a radical pace.

The experience of everyday organisational power within a cocoon of basic privilege is much more positive than it is in more exposed and bleaker situations. Indeed we have focused on such everyday practices in minute detail in this book: the dual career track, the loser-recovery system, separate job and status ladders with many rungs and small differentials, and the explicitly gendered world of work – all represent the positive use of management power. The use of negative, coercive power within these companies is simply not an option. Consensus is built slowly, carefully, and maintained by a cohort of people who will spend about 35 years in the same company – and in the company of each other. Sometimes, as doubtless seems to be the case to critical observers, the atmosphere and culture might seem overwhelmingly patriarchal and stifling, but the innovative results are remarkable, as our book attests.

Japan is often characterised as a world of managed beauty and deceptive charm, a country where exquisite refinement is an everyday accomplishment, in which great attention is paid to the presentation of oneself as just one among many others. These characterisations are accurate and give rise to a specific organisational dynamics in which excessive egoism, heroic individualism and the struggles attendant upon its realisation are the exceptions rather than the rule. Thus, as we have sought to capture in our book, within the big name companies these dynamics are embedded in disciplinary practices that inscribe action that is fluid, committed and exceptionally well-managed from the perspective of innovative organisations.

Our contention is that would-be innovative organisations elsewhere in the world should not ignore the innovation and learning that is still occurring in Japan. It can be translated. It will alter as it is translated, it will assume local characteristics and it may not work as well as it does in Japan, but it need not be alien or surreal. Our book is thus an inducement to translation.

Appendix 1: Types of Strategic Decisions

Decision process	*Innovative and analytical (planning mode)*	*Innovative and intuitive (entrepreneurial mode)*	*Conservative and bureaucratic (bureaucratic mode)*	*Conservative and intuitive (reactive mode)*
Organisational level of planning	Planning staff and top management	Top-down	Relatively bottom-up	(Miscellaneous)
Goal and policy	Clear goals, high goals	Unclear goals, value in innovation	Idealistic and perfectionist, 'safety first'	Unclear goals, value in safety
Information collection and idea generation	From information to ideas, sensitive to new information	Intuitive ideas, sensitive to new opportunities	Reluctant to make decisions until sure of enough information, avoid uncertainty	Ideas by hunch and experience
Search	Aggressive research, many alternatives	Aggressive and adventurous search, few alternatives	Conservative search, few alternatives	When problems happen
Time horizon	Long (far ahead of competition)	Long	Short	Short (a follower)
Evaluation	Deliberate evaluation, considers the worst case	Adventurous quick decisions, without considering resources	Do not cross the stone bridge, suboptimisation	Avoid risk
Integration	Integration by comprehensive planning	Build-up approach, first-in, first-out	Integration is done within the limits of present resources	Build-up approach
Size of move	Innovative and large	Innovative and large	Incremental	Incremental or imitative
Fitting decisions	Large projects	Small projects	–	–
Fitting departments	Top management, planning department	Marketing department	Personnel department, finance department	
Cases	Canon, Hitachi, Sony	Nintendo, Secom, Kyosera, Yamaha (in early stages). Some companies bankrupted by being too aggressive	Former Japanese National Railways, government enterprises	Van Jacket Co., some camera makers

Note: These classifications are based on observations of actual cases and an analysis of the literature, including Mintzberg (1973), Ansoff *et al.* (1976) and Miles and Snow (1978).

Appendix 2: Companies Responding to the Mail Survey, 1995

Industry	*Number of responding companies*
Construction	3
Food	6
Textiles	7
Paper	3
Chemicals and drugs	12
Oil refining	2
Rubber	4
Glass and cement	3
Steel, non-ferrous metals	9
Machinery	9
Electrical machinery, precision machinery	20
Transportation equipment	13
Other	6
Total	97

Notes: Thirty-three of the companies were specialised, 64 were diversified. The survey was mailed to 200 manufacturing companies listed on the Tokyo stock exchange. The survey items addressed the organisational structure and long-range planning system of the company.

Appendix 3: The Functions of the Head Office in Successful Japanese Companies

The development of corporate strategy

- Planning:
  - promoting strategic thinking,
  - collecting strategic information,
  - presenting new strategies,
  - integrating strategies across divisions and functions.
- Research and development management:
  - technology forecasting, long-term and short-term,
  - planning longer-term research that is beyond the scope of divisions,
  - organising research activities, acquiring resources and allocating them,
  - Developing project teams, establishing incubator departments.
- Development of new products and new business:
  - improving the product development process,
  - evaluating development plan,
  - improving interface between development, production and marketing,
  - promoting new product development,
  - as an incubator department, developing many infant projects.
- Environmental protection:
  - promoting environmental protection programme.
- Supervision of subsidiaries;
  - determining the mission of subsidiaries,
  - planning the supporting activities of the head office, including personnel management,
  - integrating activities across subsidiaries.
- Multinational management:
  - evaluating new business opportunities,
  - integrating regional activities,
  - integrating marketing,
  - coordinating personnel management.
- Financial control:
  - budgetary control and capital budgeting,
  - developing new accounting systems.

Expert staff assistance to develop strong core competencies

- Personnel department:
  - centralised recruitment of university graduates,

  - establishing the personnel management system,
  - coordinating personnel management.
- Production technology and quality assurance:
  - developing production systems, such as JIT,
  - promoting quality control.
- Marketing:
  - research into marketing and customer service,
  - marketing research and competitive benchmarking,
  - developing new marketing channels, e.g. marketing on the internet.

Centralised services to produce strong competencies

- Personnel services:
  - planning employee training,
  - welfare system.
- Finance:
  - raising funds,
  - financial control,
  - relations with financial institutions.
- Legal:
  - corporate legal issues,
  - patents and copyrights.
- Procurement and logistics:
  - purchasing (in domestic and international markets),
  - logistic systems.
- Management of capital investment:
  - planning capital investment,
  - construction management,
  - maintenance,
  - property management.
- Information system:
  - developing information system,
  - operating centralised information systems.
- Corporate communications:
  - public relations,
  - advertising and sales promotion,
  - government relations.
- Marketing department and production department where the organization has a functional organisational structure.

Bibliography and Suggestions for Further Reading

AAKER, D. A. (1984) *Strategic Market Management* (New York: John Wiley).
ABEGGLEN, J. C. (1958) *The Japanese Factory: Aspects of Its Social Organization* (Glencoe, Ill.: Free Press).
ABEGGLEN, J. C. and G. STALK (1985) *Kaisha: The Japanese Corporation* (New York: Harper & Row).
ABEL, D. E. and J. S. HAMMOND (1979) *Strategic Market Planning* (New York: Prentice-Hall).
ABELL, D. F. (1980) *Defining the Business? The Starting Point of Strategic Planning* (New York: Prentice-Hall).
AKIYAMA, T. (1962) *Sozo no Shinri* (Psychology of Creativity) (Tokyo: Seishin Shobo).
ALLEN, T. J. (1977) *Managing the Flow of Technology* (Cambridge, Mass.: MIT Press).
American Bar Association (1994) *Corporate Director's Guidebook* (Chicago: ABA).
AMIT, R. and P. H. SHOEMAKER (1993) 'Strategic Assets and Organization Rent', *Strategic Management Journal*, January, pp. 33–46.
ANDREWS, K. R. (1971) *The Concept of Corporate Strategy* (New York: Dow Jones-Irwin).
ANSOFF, H. I. (1965) *Corporate Strategy* (New York: McGraw-Hill).
ANSOFF, H. I., R. P. DECLERCK and R. L. HAYES (eds) (1976) *From Strategic Planning to Strategic Management* (London: John Wiley).
ANSOFF, H. I. and Stewart J. (1967) 'Strategies for a Technology-based Business', *Harvard Business Review*, November–December, pp. 71–83.
AONUMA, Y. (1965) *Nihon no Keieiso* (Japanese Top Management) (Tokyo: Nihon Keizai Shinbun Sha).
ARGYRIS, C. (1964) *Integrating the Individual and the Organization* (New York: John Wiley).
Asahi Newspaper (ed.) (1999) *Japan Almanac* (Tokyo: Asahi Newspaper).
BADEN-FULLER, C. and J. M. STOPFORD (1988) 'Why Global Manufacturing?', in H. V. WORTZEL and L. H. WORTZEL (eds), Global Strategic Management (New York: John Wiley), pp. 391–9.
BADEN-FULLER, C. and J. M. STOPFORD (1994) *Rejuvenating Mature Business* Boston, Mass.: Harvard University Press).
BAIN, J. S. (1959, 1968) *Industrial Organization* (New York: John Wiley).
BARBAR, A. E. and C. L. DALY (1996) 'Compensation and Diversity, New Pay for a New Workforce?', in E. E. KOSSEK, and S. A. LOBEL, *Managing Director* (Oxford: Blackwell), pp. 194–216.
BARNARD, C. I. (1938) *The Functions of the Executive* (Cambridge, Mass.: Harvard University Press).

BARTLETT, C. A., Y. DOZ and G. HEDLUND (eds) (1990) *Managing the Global Firm*, (New York: Routledge).

BARTLETT, C. A. and S. GHOSHAL (1986) 'Tap Your Subsidiaries for Global Reach', *Harvard Business Review*, November–December, pp. 87–94.

BARTLETT, C. A. and S. GHOSHAL (1994) 'Changing the Role of Top Management: Beyond Strategy to Purpose' *Harvard Business Review*, November–December, pp. 79–88.

BARTLETT, C. A. and H. YOSHIHARA (1988) 'New Challenges for Japanese Multinationals', in V. PUCIK, N. M. TICHY, and C. K. BARNET (eds) *Globalizing Management* (New York: John Wiley), pp. 276–99.

BASS, B. M. (1985) *Leadership and Performance Beyond Expectation* (New York: Free Press).

BEACH, D. S. (1985) *Personnel* (New York: Macmillan).

BEAMISH, P. W. and A. C. INKPEN (1995) 'Keeping International Joint Ventures Stable and Profitable', *Long Range Planning*, June, pp. 26–36.

BEAUMONT, P. B. (1993) *Human Resource Management, Key Concepts and Skill* (London: Sage).

BECKER, B. and B. GERHART (1996) 'The Impact of Human Resource Management on Organizational Performance: Progress and Prospects', *The Academy of Management Journal*, August, pp. 779–801.

BENEDICT, R. (1946) *The Chrysanthemum and the Sword* (New York: Houghton Mifflin).

BERLE, A. A. and G. C. MEANS (1932/1967) *The Modern Corporation and Private Property* (New York: Macmillan).

BIRD, F. B. and J. WALTERS (1989) 'The Moral Muteness of Managers', *California Management Review*, Fall, pp. 74–88.

BLACKBURN, R. and B. ROSEN (1993) 'Total Quality and Human Resources Management: Lessons learned from Baldrige Award winning companies', *Executive*, vol. 7, no. 3 pp. 49–66.

BLAKE, R. R. and J. S. MOUTON (1964) *The Managerial Grid* (Houston, TX: Gulf Publishing).

BLAU, P. M. (1964) *Exchange and Power in Social Life* (New York: John Wiley).

BLEEKE, J. and D. ERNST (1991) 'The Way to Win in Cross-border Alliances', *Harvard Business Review*, November–December, pp. 127–35.

BLEEKE, J. and D. ERNST (1995) 'Is Your Strategic Alliance Really a Sale?', *Harvard Business Review*, January–February, pp. 97–105.

BOLT, J. K. (1988) 'Global Competitors: Some Criteria for Success', *Business Horizon*, January–February.

BRANDENBURGER, A. M. and B. J. NALEBUFF (1995) 'The Right Games: Use Game Theory To Shape Strategy', *Harvard Business Review*, July–August, pp. 57–73.

BRYMAN, A. (1996) 'Leadership in Organizations', in S. R. CLEGG, C. HARDY and W. R. NORD (eds), *The Handbook of Organization Studies* (London: Sage), pp. 276–91.

BUCKLEY, A. (1975) Growth by Acquisition, *Long Range Planning*, August, pp. 53–60.

BURCH, P. H. (1972) *The Managerial Revolution Reassessed* (Boston, Massachusetts: D. C. Heath).

BURNHAM, J. (1941) *Managerial Revolution* (New York: John Day).
BURNS, T. and G. M. STALKER (1961) *The Management of Innovation* (London: Tavistock).
BUSHELE, R. (1962) 'How to Evaluate a Firm', *California Management Review*, Fall, pp. 5–17.
BUZZEL, R. D. and B. T. GALE (1987) *The PIMS Principles* (New York: The Free Press).
CALORI, R. (1988) 'How Successful Companies Manage Diverse Businesses', *Long Range Planning*, vol. 21, no. 3, pp. 80–9.
CANE, S. (1997) *Kaizen Strategies – For Winning through People* (Washington, DC: Pitman).
CAVALERI, S. and D. FEARON (eds) (1996) *Managing in Organizations That Learn* (Oxford: Blackwell).
CAVES, R. (1964, 1966) *American Industry: Structure Conduct and Performance* (Englewood Cliffs, NJ: Prentice-Hall).
CHAMBERLIN, E. (1933) *The Theory of Monopolistic Competition* (Cambridge, Mass.: Harvard University Press).
CHAMPY, J. (1995) *Reengineering Management, The Mandate for New Leadership* (New York: Harper Business).
CHANDLER, A. D. Jr (1962) *Strategy and Structure*, Boston, MIT Press.
CHANDLER, A. D. Jr (1991) 'The Functions of the HQ Unit in the Multibusiness Firm', *Strategic Management Journal*, Winter, pp. 31–50.
CHADWICK-JOHNS, J. K. (1976) *Social Exchange Theory* (London: Academic Press).
CHARKHAM, J. (1994) *Keeping Good Company* (New York: Oxford University Press).
CHOKKI, T. (1993) *Gendai no Keieisoshiki* (Business Organization), (Tokyo: Yuhikaku).
CLARK, R. D. (1971) 'Group-induced Shift Toward Risk, A Critical Appraisal', *Psychological Bulletin*, vol. 76, no. 4.
CLARKE, T. and S. CLEGG (1998) *Changing Paradigms* (London: HarperCollins).
CLEGG, S. R., D. DUNPHY and G. REDDING (1986) *The Enterprise and Management in East Asia*, Centre of Asian Studies, University of Hong Kong Press.
CLEGG, S. R. and S. G. REDDING (eds) (1990) *Capitalism in Contrasting Cultures*, Berlin, De Gruyter.
CLEGG, S. R. (1990) *Modern Organizations: Organization Studies in the Postmodern World*, London, Sage.
CLEGG, S. R. (1994) 'Power Relations and the Constitution of the Resistant Subject', in J. M. JERMIER, D. KNIGHTS and W. R. NORD (eds), *Resistance and Power in Organizations* (London: Roultledge), pp. 274–325.
COLLINS, J. C. AND J. I. PORRAS (1994) *Built to Last: Successful Habits of Visionary Companies* (New York: Curtis Brown).
COLLIS, D. J. and C. A. MONTGOMERY (1995) 'Competing on Resources: Strategy in the 1990s', *Harvard Business Review*, July–August, pp. 118–28.
CONTRACTOR, F. J. and P. LORANGE (eds) (1988) *Cooperative Strategies in International Business* (Massachusetts: Lexington Books).

CRAWFORD, R. J. (1998) *Reinterpreting the Japanese Economic Miracle* (Cambridge, Mass.: Harvard, University Press).

CYERT, R. and J. MARCH (1963) *A Behavioral Theory of the Firm* (Englewood Cliffs, NJ: Prentice-Hall).

D'AVENI, R. A. (1994) *Hypercompetition: Managing the Dynamics of Strategic Manoeuvring* (New York: Free Press).

D'AVENI, R. A. (1995a) *Hyper Competitive Rivalries* (New York: Free Press).

D'AVENI, R. A. (1995b) Coping with Hypercompetition: Utilizing the New 7Ss Framework', *Academy of Management Executive*, vol 9, no. 3, pp. 45–60.

DAVIES, S. M. and P. R. LAWRENCE (1977) *Matrix Organization* (Reading, Mass.: Addision-Wesley).

DEAL, T. E. and A. A. KENNEDY (1982) *Corporate Culture* (Reading, Mass.: Addison-Wesley).

DEAN, J. (1951) *Managerial Economics* (Englewood Cliffs, NJ: Prentice-Hall).

DE BONO, E. (1967) *New Think* (New York: Charles E. Tuttle).

DELERY, J. E and D. H. DOTY (1996) 'Modes of Theorizing in Strategic Human Resource Management. Tests of Universalistic Contingency and Configurational Performance Predictions', *The Academy of Management Journal*, vol. 39, no. 4, pp. 802–35.

DENTSU SOKEN (Dentsu Institute of Human Studies) and Leisure Development Center (1995) *Global Survey of Human Values* (Tokyo: Dentsu Soken).

DEUTCH, M. (1969) 'Conflicts: Productive and Destructive', *Journal of Social Issues*, January.

DEUTCH, M. (1973) *The Resolution of Conflict* (New Haven, CT: Yale University Press).

DOI, K. (1971) *Amae no Kozo* (Structure of Reliance) (Tokyo: Kobundo).

DONALDSON, L. (1999) *Performance-Driven Organizational Change: The Organizational Portfolio* (London: Sage)

DORE, R. (1973) *British Factory–Japanese Factory: The Origins of National Diversity in Industrial Relations* (Berkeley, CA: University of California Press).

DOZ, Y. L. (1986) *Strategic Management in Multinational Companies* (Oxford: Pergamon Press).

DREW, S. A. W. (1997) 'From Knowledge to Action: The Impact of Benchmarking on Organizational Performance', *Long Range Planning*, June, pp. 427–41.

DRUCKER, P. (1974) *Management; Tasks, Responsibilities, Practices* (New York Harper & Row).

DRUCKER, P. F. (1954) *The Practice of Management* (New York: Harper & Brothers).

DUNNING, J. H. (1988) *Explaining International Production* (London: HarperCollins).

DUNNING, J. H. (1993) *Multinational Enterprises and the Global Economy* (Wokingham: Addison-Wesley).

DUNPHY, D. (1986) 'An Historical Review of the Literature on the Japanese Enterprise and its Management', S. R. CLEGG, D. DUNPHY and S. G. REDDING (eds), *The Enterprise and Management in East Asia* (Hong Kong: Center of Asian Studies, University of Hong Kong).

DUNPHY, D. (1987) 'Convergence/Divergence: A Temporal Review of the Japanese Enterprise and its Management', *Academy of Management Review*, vol. 12, no. 3, pp. 445–59.

DYER, J. H. (1995) 'Improving Performance by Transforming Arms-Length Relationships to Supplier Partnerships; The Chrysler Cases', paper presented at the 1995 Annual Conference of the Academy of Management, Vancouver.

DYER, J. H. (1996a) 'How Chrysler Created an American Keiretsu', *Harvard Business Review*, vol. 74, no. 4, pp. 42–6.

DYER, J. H. (1996b) 'Specialized Suppliers Networks as a Source of Competitive Advantage: Evidence from the Auto Industry', *Strategic Management Journal*, April, pp. 271–91.

ECCLES, R. G. (1985) *The Transfer Pricing Decision* (Lexington, Mass.: Lexington Books).

ETZIONI, A. (1961) *A Comparative Analysis of Complex Organizations* (New York: Free Press).

FAIR TRADE COMMISSION (1992) *Kosei Torihiti imkai nenji Hokokusho* (Annual Report of Fair Trade Commission) Tokyo: Fair Trade Commission.

FAIR TRADE COMMISSION (1997) *Kosei Torihiki Iinkai Nenji Hokokusho* (Annual Report of Fair Trade Commission) (Tokyo: Fair Trade Commission).

FARMER, R. and B. M. RICHMAN (1965) *Competitive Management and Economic Progress* (Illinois: Richard Irwin).

FAYERWEATHER, J. (1982) *International Business Strategy and Administration* (London: Ballinger).

FIELDER, F. E. (1967) *A Theory of Leadership Effectiveness* (New York: McGraw-Hill).

FLAHERTY, M. T. (1996) *Global Operations Management* (New York: McGraw-Hill).

GALBRAITH, J. K. (1967) *The New Industrial State* (Boston, Mass.: Houghton Mifflin).

GALBRAITH, J. K. (1969) *The Affluent Society* (Boston, Mass.: Houghton Mifflin).

GALBRAITH, J. K. and D. A. NATHANSON (1978) *Strategy Implementation, Role of Structure and Process* (Los Angeles, CA: West Publishing).

GALBRAITH, J. R. (1977) *Organization Design* (Reading, Mass.: Addison-Wesley).

GALBRAITH, J. R. (1984) 'Designing the Innovation Organization', in R. B. LAMB (ed.), *Competitive Strategic Management* (Englewood Cliffs, NJ: Prentice-Hall).

GARVIN, D. A. (1993) 'Building a Learning Organization', *Harvard Business Review*, July–August, pp. 78–92.

GHOSHAL, S. and C. A. BARTLETT (1990) 'The Multinational Corporation as an Interorganizational Network', *Academy of Management Review*, 15 October, pp. 603–25.

GHOSHAL, S. and C. A. BARTLETT (1995) 'Changing the Role of Top Management: Beyond Structure to Processes', *Harvard Business Review*, Vol. 73, no. 1, pp. 86–96.

GHOSHAL, S. and N. NOHRIA (1993) 'Horses for Courses: Organizational Forms for Multinational Corporations', *Sloan Management Review*, vol. 34, no. 2, pp. 23–35.

GHOSHAL, S. and N. NOHRIA (1995) 'Horses for Courses', in J. DREW (ed), *Readings in International Enterprise* (London: Routledge) p. 49.

GIBSON, J. L., J. M. IVANCEVICH and J. H. DONNELY (1988) *Organizations: Behavior, Structure and Business* (Irwin, Ill.: BPI).

GLUECK, W. E. (1976) *Business Policy, Strategy Formulation and Management Action* (New York: McGraw-Hill).

GOOLD, M. A. CAMPBELL and M. ALEXANDER (1994) *Corporate Level Strategy, Creating Value in a Multibusiness Company* (New York: John Wiley).

GORDON, R. A. (1945) *Business Leadership in the Large Corporation* (Washington, DC: Brookings Institution).

GORT, M. (1962) *Diversification and Integration in American Industry* (Princeton, NJ: Princeton University Press).

GOSS, T., R. PASCALE and A. ATHOS (1992) (1993) 'The Reinvention Roller Coaster: Risking the present for a powerful future', *Harvard Business Review*, vol. 71, no. 6, pp. 97–108.

GRANICK, D. (1962) *The European Executive* (London: Doubleday).

GRANT, J. L. (1997) *Foundations of Economic Value Added* (Pennsylvania: Habozzi Associates).

GREENLEY, G. E. (1989) *Strategic Management* (London: Prentice-Hall).

GREINER, L. E. (1972) 'Evolution and Revolution as Organizations Grow', *Harvard Business Review*, July–August, pp. 37–46.

GROENEWEGEN, J. (1997) 'Institutions of Capitalism: American, European and Japanese Systems Compared', *Journal of Economic Issues*, vol. 31, no. 2 pp. 333–47.

GUTENBERG, E. (1956) *Grundlagen de Batriebswirtschaftslehre, Der Absatz* (Berlin: Springer-Verlez).

GUTH, W. D. and R. TAGIURI (1965) 'Personal Values and Corporate Strategies, *Harvard Business Review*, vol. 43, pp. 123–32.

GUTKNECHT, J. E. and J. B. KEYS (1993) 'Mergers, Acquisition and Takeovers: Maintaining Morale of Survivors and Protecting Employees', *Executive*, August.

HAEFELE, J. H. (1962) *Creativity and Innovation* (New York: Reinhold).

HAGEDOORN J. (1993) 'Understanding The Rationale of Strategic Technology Partnering: Interorganizational modes of cooperation and sectoral differences', *Strategic Management Journal*, July, pp. 371–85.

HAIGH, R. W. (1992) 'Building a Strategic Alliance, The Hermosllo Experience as a Ford Mazda Proving Ground', in M. T. FLAHERTY, *Global Operations Management*, (New York: McGraw-Hill).

HALL R. (1992) 'The Strategic Analysis of intangible resources', *Strategic Management Journal*, vol. 13, no. 2, pp. 135–44.

HALL, R. (1993) 'A Framework Linking Intangible Resources and Capabilities to Sustainable Competitive Advantage', *Strategic Management Journal*, vol. 14, no. 8, pp. 607–18.

HALL R. H. (1982) *Organizations, Structure and Process* (Englewood Cliffs, NJ: Prentice-Hall).

HALL, R. J. and C. J. HITCH (1939) *Price Theory and Business Behavior*, Oxford Economic Papers.

HALL, W. K. (1980) 'Survival Strategies in a Hostile Environment', *Harvard Business Review*, vol. 58, no. 5, pp. 75–85.

HAMEL, G. and C. K. PRAHALAD (1989) 'Strategic Intent', *Harvard Business Review*, vol. 67, no. 3, pp. 63–76.

HAMEL, G. and C. K. PRAHALAD (1994) *Competing for the Future* (Cambridge, Mass.: Harvard Business School).

HAMMER, M. and J. CHAMPY (1993) *Re-engineering the Corporation* (New York: Free Press).

HARBISON, F. (1959) 'Management in Japan', in F. HARBISON and C. A. MYERS (eds), *Managing in the Industrial World: An International Analysis* (New York: McGraw-Hill).

HARRIGAN, K. (1985) *Strategies for Joint Ventures* (Lexington, Mass.: Lexington Books).

HARRIGAN, K. R. (1986) *Managing for Joint Venture Success* (Lexington, Mass.: Lexington Books).

HAYES, R. H. and W. J. ABERNATHY (1980) 'Managing Our Way to Economic Decline', *Harvard Business Review*, July–August, pp. 67–77.

HAZAMA, H. (1976) *Nihonteki Keiei no Keifu* (Origin of the Japanese Management System) (Tokyo: Nihon Noritsukyokai).

HENDERSON, B. D. (1979) *Henderson on Corporate Strategy* (Massachusetts: Abt Books).

HENDERSON, J. M. and R. E. QUANDT, (1958) *Microeconomic Theory: A Mathematical Approach* (New York: McGraw-Hill).

HERSEY, P. and K. H. BLANCHARD (1993) *The Management of Organizational Behavior* (Englewood Cliffs, NJ: Prentice-Hall).

HILL, C. and R. HORKINSON (1987) 'Strategy and Structure in the Multiproduct Firm', *Academy of Management Review*, vol. 12, no. 2, pp. 331–41.

HIRABAYASHI, Y. and T. SASA (1996) *Nyumon ISO14,000* (Introduction to ISO 14,000) (Tokyo: Nikkagiren).

HIRATA, S. (1989) *Sony Tsuyosa no Himitsu* (Secret of Strength of Sony) (Tokyo: OS Shuppan).

HITT, M. A. and R. D. IRELAND (1985) 'Corporate Distinctive Competence, Strategy, Industry and Performance', *Strategic Management Journal*, vol. 6, no. 3, pp. 273–93.

HODGE, B. J., W. P. ANTHONY and L. M. GALES (1996) *Organization Theory, A Strategic Approach* (Englewood Cliffs, NJ: Prentice-Hall).

HOFER, C. W. and D. SCHENDEL (1987) *Strategy Formation: Analytical Concepts* (West Pub. Co: St Paul's, Minnesota).

HOFSTEDE, G. (1980) *Culture's Consequences: International Differences in Work-Related Values* (California: Sage).

HOLDEN, P. E., L. S. FISH and H. L. SMITH (1941) *Top Management Organization and Control* (Stanford, CA: Stanford University Press).

HOSKISSON, R. (1987) 'Multidivisional Structure and Performance: the contingency of diversification strategy', *Academy of Management Journal*, vol. 30, no. 4, pp. 625–44.

HOSKISSON R. E., M. A. HITT and R. A. JOHNSON (1993) 'Construct Validity of an Objective Categorical Measure of Diversification Strategy', *Strategic Management Journal*, March, pp. 215–35.

HOSMER, L. T. (1995) 'Trust: The Connecting Link Between Organizational Theory and Philosophical Ethics', *Academy of Management Review*, vol. 20, no. 2, pp. 379–403.

HOUSE, R. J. (1976) 'A Theory of Charismatic Leadership', in J. G. HUNT and L. L. LARSON, *Leadership: The Cutting Edge* (Southern Illinois University Press).

IIDA, F. (1998) *Nihon – teki Keiei no Ronten* (Review of Japanese Management Studies) (Tokyo: PHP).

ISHIDA, H. (ed.) (1984) *Kokusai Keiei no Ningen Mondai* (Problems of Human Resource Management in Multinational Management) (Tokyo: Keio Tsushin).

ISHIDA, H. (1985) *Nihon Kigyo no Kokusai Jinjikanri* (Personnel Management of Japanese Foreign Subsidiaries) (Tokyo: Nihon Rodo Kyokai).

ISHII, K. (1996) *Gendai Nihon Kigyo no Keieisha* (Top Management of Japanese Corporations (Tokyo: Bunshindo).

ISHII, J., A., OKUMURA, T. KAGONO and I. NONAKA (1996) *Keieisenruaku-Ron* (Corporate Strategy) (Tokyo: Yuhikaku).

ISHIKURA, K. (1988) 'Achieving Japanese Productivity and Quality Levels at a U.S. Plant', *Long Range Planning*, vol. 21, no. 5, pp. 10–17.

ITAMI, T. (1984) *Shin Keieisenryaku no Ronri* (New Principles of Corporate Strategy), (Tokyo: Nihon Keizai).

ITAMI, T., T., KAGONO, M. MIYAMOTO, and S. YONEKURA (1997) *Nihon-teki Keiei no Seisei to Hatten* (Development of the Japanese Management System) (Tokyo: Yuhikaku).

IZAWA, Y. (1996) *Nihon Kigyo Global-ka no Kouzu* (Structure of Japanese Multinational Management) (Tokyo: Gakubun-sha).

JAPAN PRODUCTIVITY CENTER (ed.) (1986) *Shaze Shakun* (Business Creeds) (Tokyo: JPC).

JAPAN PRODUCTIVITY CENTER (ed.) (1998) *Shaze Shakun* (Business Creeds) (Tokyo: JPC).

KAGAKU GIJUTSU CHO (1995) *Kagaku Gijutsu Hakusho* (White Paper on Science and Technology) (Tokyo: Ministry of Science and Technology).

KAGANO, T. (1997) *Nihongata Keiei no Fukken* (Recovery of Japanese Management), (Tokyo: PHP).

KANAI, T. (1991) *Henkaku-gata Middle no Tankyu* (In Search of Innovative Middle Managers), (Tokyo: Hakuto Shobo).

KANTER, R. M. (1983) *The Change Masters* (New York: Simon & Schuster).

KANTER, R. M. (1994) 'Collaborative Advantage: The Art of Alliances', *Harvard Business Review*, July–August, pp. 96–108.

KAPLAN, R. S. and D. P. NORTON (1992) 'The Balanced Scorecard: Measures that Drive Performance', *Harvard Business Review*, vol. 70, no. 1, pp. 71–9.

KAST, F. E. and J. E. ROSENZWEIG (1973) *Contingency Views of Organization and Management* (Chicago: Science Research Associates).

KATZ, D. and R. L. KAHN (1966) *The Social Psychology of Organization* (New York: John Wiley).

KATZ, R. L. (1955) 'Skills of an Effective Administrator', *Harvard Business Review*, vol. 52, no. 5, pp. 33–42.

KAWAKITA, J. (1967) *Hassoho* (New Idea Generation) (Tokyo: Chuokoron).

KEIZAI KIKAKUCHO (ECONOMIC PLANING BUREAU) (1991) *Kokumin no Ishiki to Needs* (National Attitudes and Needs) (Tokyo: Keizai Kikakucho).

KEPNER, C. H. and B. B. TREGOR (1965) *The Rational Manager, A Systematic Approach to Problem Solving and Decision Making* (New York: McGraw-Hill).

KIMURA, N. (1976) *Tasogare no London kara* (Evening Twilight in London) (Tokyo: PHP).

KITCHING, J. (1967) 'Why Do Mergers Miscarry?', *Harvard Business Review*, vol. 45, pp. 84–101.

KOBAYASHI, N. (1980) *Nihon no Kokusai Kigyo* (Japanese Multinational Management) (Tokyo: Chuokeizai).

KOJIMA, K. (1989) *Kaigai Chokusetsu Toshi no Macro Bunseki* (Macro Analysis of Foreign Direct Investment) (Tokyo: Bunshindo).

KOJIMA, I. (1998) *Nihon no Kigyo Group* (Japanese Business Group) (Tokyo: Paru Shuppan).

KOMIYA, K. (1998) *Cash Flow Keiei* (Cash Flow Management) (Tokyo: Toyokeizai).

KONDO, S. (1985) *Shinseihin, Shinjigyo Tansaku Ho* (Exploration of New Products and New Business) (Tokyo: Nihon Noritsu Kyokai).

KONO, T. (1966) *Keiei Keikaku no Riron* (Theory of Business Planning) (Tokyo: Diamond Sha).

KONO, T. (1971) *Keiei Hoshin Ron* (Business Policy) (Tokyo: Hakuto Shobo).

KONO, T. (1974) *Keiei Senryaku no Kaimei* (Analysis of Corporate Strategy) (Tokyo: Diamond-sha).

KONO, T. (1976) Long-range Planning, Japan–USA: A Comparative Study, *Long Range Planning*, vol. 9, no. 5, pp. 61–71.

KONO, T. (1984a) *Strategy and Structure of Japanese Enterprises* (London: Macmillan).

KONO, T. (1984b) 'Asia ni okeru Nihon no Takokuseki Kigyo, no. 3' (Japanese Multinational Companies in Asia, no. 3), *Gakushuin Daigaku Keizai Ronshu*, March and September, 1983.

KONO, T. (1987a) *Keieigaku Genri* (Principles of Management) (Tokyo: Hakuto-Shobo).

KONO, T. (1987b) *Shin Seihin Kaihatsu Senryaku* (New Product Development) (Tokyo: Diamond Sha).

KONO, T. (1988) 'Factors Affecting the Creativity of Organizations – An Approach from the Analysis of New Product Development', in K. URABE, J. CHILD and T. KAGONO (eds), *Innovation and Management* (Berlin: Water De Gruyter), pp. 105–44.

KONO, T. (1992a) 'Organizational Problems of Research and Development', *International Journal of Technology Management*, vol. 7, nos 1–3, pp. 61-76.

KONO, T. (1992b) *Long-Range Planning of Japanese Corporations* (Berlin: Walter de Gruyter).

KONO, T. (1992c) *Strategic Management in Japanese Companies* (Oxford: Pergamon Press)

KONO, T. (1999) *Shin Gendai no Keiei Senryaku* (Modern Corporate Strategy, new edition) (Tokyo: Diamond).

KONO, T. and S. CLEGG (1998) *Transformation of Corporate Culture* (Berlin: Walter de Gruyter).

KOONTZ, H. (1967) *The Board of Directors and Effective Management* (New York: McGraw-Hill).

KOONTZ, H. and C. J. O'DONNEL (1959) *Principles of Management* (New York: McGraw-Hill).

KOTLER, P. (1980, 1988) *Marketing Management*, 4th edn (Englewood Cliffs, NJ: Prentice-Hall).

KOTTER, J. P. and J. L. HESKETT (1992) *Corporate Culture and Performance* (New York: Free Press).

KRIESBERG, L. (1973) *The Sociology of Social Conflicts* (New York: Prentice-Hall).

KUNITOMO, R. (1993) *Seven Eleven no Keiei Kakumei* (Innovation of Seven Eleven) (Tokyo: Parushuppan).

KUROKAWA, A. (1997) 'Global Strategy of Design and Development by Japanese Car Makers', paper presented at the 1997 Annual Meeting of the Association of Japanese Business Studies, Washington, DC.

KUWANO, Y. and K. YOSHIDA (1997) 'Sanyo's Flat Diamond: Integrating R&D into the Business', *Long Range Planning*, June, pp. 483–92.

LASSAR, W. M. (1996) Strategy and Control in Supplier–Distributor Relationships, an Agency Perspective', *Strategic Management Journal*, October, pp. 613–32.

LAWLER, E. E. and S. A. MOHRMAN (1985) 'Quality Circles after the Fad', *Harvard Business Review*, vol. 63, no. 1 pp. 64–71

LAWRENCE, P. R. and J. W. LORSCH (1967) *Organization and Environment* (Cambridge, Mass.: Harvard University Press).

LEI, D. (1993) 'Offensive and Defensive Uses of Strategic Alliances', *Long Range Planning*, August.

LERNER, A. P. (1934) 'The Concept of Monopoly Power', *Review of Economic Studies*, vol. 1.

LEVITT, T. (1960) 'Marketing Myopia', *Harvard Business Review*, July–August, pp. 45–56

LEVITT, T. (1963) 'Creativity is not Enough', *Harvard Business Review*, May–June, pp. 72–83.

LIKERT, R. (1967) *The Human Organization* (New York: McGraw-Hill).

LIKERT, R. (1969) *New Patterns of Management* (New York: McGraw-Hill).

LINDBLOM, C. E. (1968) *The Policy-Making Process* (Englewood Cliffs, NJ: Prentice-Hall).

LINOUWES, R. G. (1993) 'The Japanese Manager's Traumatic Entry into the United States: Understanding the American–Japanese Cultural Divide, Executive Commentary', *The Academy of Management Executive*, vol. 7, no. 4, pp. 21–40.

LLOYD, B. (ed.) (1977) *Creating Value Through Acquisitions, Demergers, Buyouts and Alliances* (Oxford: Pergamon).

LORANGE, P. and I. ROSS (1992) *Strategic Alliances* (Massachusetts: Blackwell).

LOUCKS, V. (1987) 'A CEO Looks at Ethics', *Business Horizons*, vol. 30, no. 2, pp. 2–6.

MCGREGOR, D. (1960) *The Human Side of Enterprise* (New York: McGraw-Hill).

MACLELLAND, D. C. (1966) 'That Urge to Achieve', in D. A. KOLB, I. M. RUBIN and J. M. MCLINTYRE (eds), *Organizational Psychology*, (Englewood Cliffs, NJ: Prentice-Hall), pp. 147–54.

MAKADOC, R. (1995) 'Distinguishing and Measuring Six Different Forms of First-Mover Advantage for Product Innovations', paper presented to the American Academy of Management Annual Meeting, Vancouver.

MAKADOK, R. (1998) 'Can First-Mover and Early-Mover Advantage be Sustained in an Industry with Low Barriers to Entry/Imitation?', *Strategic Management Journal*, vol. 19, no. 7, pp. 683–96.

MALLORY, G., T. CLARK and D. PUGH (1994) 'Organizational Structure, Strategy and Structure Change in European Manufacturing Organizations, 1984', paper presented to the Strategic Management Society Conference in Paris.

MARCH, J. G. and H. A. SIMON (1958) *Organizations* (New York: John Wiley).

MARKIDES, C. C. and P. J. WILLIAMSON (1996) 'Corporate Diversification and Organization Structure', *Academy of Management Journal*, no. 2 (April), pp. 340–67.

MARKOWITZ, H. (1959) *Portfolio Selection* (New York: John Wiley).

MARSH, R. H. and H. MANNARI (1971) 'Lifetime Commitment in Japan: Roles, Norms and Values', *American Journal of Society*, vol. 76, pp. 795–812.

MARSH, R. H. and H. MANNARI (1976) *Modernization and the Japanese Factory* (Princeton, NJ: Princeton University Press).

MARX, K. (1976) *Capital, Volume 1* (Harmondsworth: Penguin).

MASAKAME, Y. (1993) 'From Pay for Job to Pay for Knowledge', *Proceedings of Annual Meeting of Japanese Academy of Management*.

MASLOW, A. H. (1970) *Motivation and Personality* (New York: Harper & Row).

MATSUMURA, S. (1988) *Kigyo Gappei, Baishu* (Merger and Acquisition) (Tokyo: Toyokeizai).

MATSUMURA, S. (1995) *Nihon no M & A* (Merger and Acquisition in Japan) (Tokyo: Chuokeizai).

MCGREGOR, E. (1985) *The Human Side of Enterprise* (New York: McGraw-Hill).

MILES, R. and C. SNOW (1978) *Organizational Strategy, Structure and Process* (New York: McGraw-Hill).

MINISTRY OF EDUCATION (1996) *Gakkuko Kihon Chosa* (Basic Survey on Schools) (Tokyo: Government Printing Office).

MINISTRY OF LABOUR (1997) *Rodo Tokei Yoron* (Handbook of Labour Statistics) (Tokyo: Ministry of Labour).

MINTZBERG, H. (1973) *The Nature of Managerial Work* (New York: Harper & Row).

MINTZBERG, H. (1978) 'Strategy Making in Three Modes', *California Management Review*, vol. 16, no. 2, pp. 44–53.

MINTZBERG, H. (1979) *The Structuring of Organization* (Englewood Cliffs, NJ: Prentice-Hall).

MINTZBERG, H. (1987) 'Crafting Strategy', *Harvard Business Review*, vol. 65, no. 4, pp. 66–75.

MINTZBERG, H. (1988) 'Strategy Making as Craft', in K. URABE, J. CHILD and T. KAGANO (eds), *Innovation and Management* (Berlin: Walter De Gruyter).

MINTZBERG, H. (1994a) 'Rethinking Strategic Planning, Part I: Pitfalls and Fallacies', *Long Range Planning*, vol. 27, no. 3, pp. 12–21.

MINTZBERG, H. (1994b) 'Rethinking Strategic Planning, Part II: New Roles for Planners', *Long Range Planning*, vol. 27, no. 3, pp. 22–30.

MINTZBERG, H. (1994c) 'The Fall and Rise of Strategic Planning', *Harvard Business Review*, vol. 72, no. 1, pp. 107–14.

MINTZBERG, H. (1994d) *The Rise and Fall of Strategic Planning* (New York: Free Press).

MITCHEL, W. and K. SINGH (1996) 'Survival of Business Using Collaborative Relationship to Commercialize Complex Goods', *Strategic Management Journal*, March, pp. 169–95.

MITI (Ministry of International Trade and Industry (various years) *Sogo Keieiryoku Shihyo* (New Indices on Managerial Capability), (Tokyo: MITI).

MITI (Ministry of International Trade and Industry) (1996) *Sogo Keieirgoku Shihyo* (Index of Corporate Strength)

MITI (1997) *Tsusho Hakusho* (White Paper on Trade) (Tokyo: Okurasho Insatsukyoku).

MITO, K., H. MASAKI and H. HARUYAMA (1973) *Dai Kigyo ni okeru Shoyu to Shihai* (Ownership and Control in Large Corporations), (Tokyo: Mirai-sha).

MOHAIR, J. and R. SPEKMAN (1994) 'Characteristics of Partnership Attributes, Communication Behavior and Conflict Resolution', *Strategic Management Journal*, vol. 15, no. 2, pp. 135–52.

MONTGOMERY, D. B. and G. L. URBAN (1969) *Management Science in Marketing* (Englewood Cliffs, NJ: Prentice-Hall).

MORIMOTO, M. (1986) *Keieisoshiki* (Business Organization) (Tokyo: Chuo-keizai).

MORIOKA, K., T. SHIOBARA and Y. HONMA (eds) (1993) *Shin Shakai-gaku Jiten* (New Encyclopedia of Sociology) (Tokyo: Yuhikaku).

MORITA, M. (1991) *Kigyo Senryaku Ron* (Corporate Strategy) (Tokyo: Shin-sei Sha).

MORRIS, G. C. (1982) *Business Economics* (London: Routledge).

MORSE, J. W. and J. J. LORSCH (1970) 'Beyond Theory', *Harvard Business Review*, May–June, pp. 61–8.

MOSCHANDREAS, M. (1994) *Business Economics* (London: Routledge).

MOUER, R. and SUGIMOTO, Y. (1986) *Images of Japanese Society: a study in the structure of social reality* London: Kegan Paul International.

MURPHY, R. T. (2000) 'Japan's Economic Crisis', *New Left Review*, vol. 2, no. 1, pp. 25–53.

MURRAY, E. A. and J. F. MOHAN (1993) 'Strategic Alliances: Gateway to the New Europe?', *Long Range Planning*, vol. 26, no. 4, pp. 102–12.

MYRDAR, G. (1988) *An Inquiry into the Poverty of Nations* (New York: Pantheon).
NAKAGAWA, K. (1972) *Keiei Rinen* (Business Philosophy) (Tokyo: Diamond Sha).
NAKAGAWA, K. (ed.) (1977) *Nihon teki Keiei* (Japanese Management) (Tokyo: Nihon Keizai Shinbun).
NAKANE, C. (1967) *Tate Shakai no Ningen Kankei* (Vertical Society) (Tokyo: Kodan-sha).
NAKANE, C. (1973) *Japanese Society* (Harmondsworth: Penguin).
NEUMAN, J. and O. MORGENSTERN (1953) *The Theory of Games and Economic Behaviour* (Princeton, NJ: Princeton University Press).
NHK (1979) *Nihonjin no Shokugyo Kan* (Japanese Attitudes Towards Jobs) (Tokyo: NHK).
NIKKO RESEARCH CENTER (1997) *Gyokai Bunseki* (Industry Analysis) (Tokyo: Keirin-shobo).
NISHIKAWA, K. (1984) *Tema Hakkutsu no Management* (Management to Find Subjects) (Tokyo: Bunshodo).
NISHIKAWA, T. (1990) *Shin Seihin Kaihatsu Program* (New Product Development Program) (Tokyo: Presiden Sha).
NODA, K. (1960) *Nihon no Juyaku* (Japanese Top Management) (Tokyo: Diamond Sha).
NONAKA, I. and H. TAKEUCHI (1995) *The Knowledge Creating Company* (New York: Oxford University Press).
NYSTROM, H. (1979) *Creativity and Innovation* (New York: John Wiley).
O'CONNOR, R. (1983) *Managing Corporate Development*, Conference Board Report 771 (London: Elsevier).
OGAWA, T. (1995) *Senryakuteki Teikei* (Strategic Alliances) (Tokyo: Elco).
OHASHI, S., S. ODA, and G. SCHANZ (eds) (1995) *Nihon-teki Keiei to Doitsu-teki Keiei* (Japanese Management and German Management) (Tokyo: Chikura-shobo).
OHINATA, Y. (1994) 'Benchmarking: the Japanese Experience', *Long Range Planning*, vol. 27, no. 4, pp. 48–53.
OKAMOTO, A. (1991) 'Creative and Innovative Research at Ricoh', *Long Range Planning*, vol. 24, no. 5, pp. 9–16.
OKAMOTO, Y. (1986) *Gendai no Keieisoshiki* (Modern Organization) (Tokyo: Nihonkeizai).
ORSSATTO, R. J. and CLEGG S. R. (1999) 'The Political Ecology of Organizations: Towards a Business Analytical Framework', *Organization and Environment* 12, 3, 263–79.
OSBORN, A. F. (1953, 1957) *Applied Imagination, Principals and Procedure of Creative Thinking* (New York: Charles Scribner).
OSTER, S. M. (1994) *Modern Competitive Analysis* (New York: Oxford University Press).
OUCHI, W. (1981) *Theory Z? How American Business Can Meet the Japanese Challenge* (Reading, MA: Addison-Wesley).
OURA, Y. (1988) *Knowledge Management Kakumei* (Knowledge Management Revolution) (Tokyo: Toyokeizai).
OYAMA, Y. (ed.) (1965) *Shaze Shakun* (Business Creeds) (Tokyo: Koyoshoin).

PARKE, A. (1993) 'Strategic Alliance Structuring', *Academy of Management Journal*, August, pp. 794–829.
PASCALE, R. T. and A. G. ATHOS (1981) *The Art of Japanese Management: Applications for American Executives* (New York: Simon & Schuster).
PEKAR, P. and R. ALLIO (1994) 'Making Alliances Work – Guidance for Success', *Long Range Planning*, vol. 27, no. 4, pp. 54–65.
PELZ, D. D. and F. M. ANDREWS (1966) *Scientists in Organizations* (New York: John Wiley).
PENROSE, E. (1959, 1966) *The Theory of the Growth of the Firm* (Oxford: Basil Blackwell).
PERLMUTTER, H. V. (1969) 'The Tortuous Evolution of the Multinational Corporation', *Columbia Journal of World Business*, January–February, pp. 9–18.
PERROW, C. (1970) *Organizational Analysis: A Sociological View* (London: Tavistock).
PERROW, C. (ed.) (1985) *Complex Organizations* (New York: McGraw-Hill).
PESSEMIER, E. A. (1977) *Product Management, Strategy and Organization* (New York: John Wiley).
PETERS, T. J. and R. H. WATERMAN (1982) *In Search of Excellence: Lessons from America's Best-run Companies* (New York: Harper & Row).
PFEFFER, J. (1981) *Power in Organizations* (Massachusetts: Pittman).
PFEFFER, J. (1998) *The Human Equation: Building Profits by Putting People First*, Boston. MA: Harvard Business School Press.
PINCHOT, G. (1985) *Intrapreneuring* (New York: Harper & Row).
PORTER, L. W. and E. E. LAWLER (1968) *Managerial Attitudes and Performance* (Homewood, Ill.: Irwin).
PORTER, M. E. (1980) *Competitive Strategy: Techniques for Analyzing Industries and Competitors* (New York: The Free Press).
PORTER, M. E. (1985) *Competitive Advantage: Creating and Sustaining Superior Performance* (New York: The Free Press).
PORTER, M. E. (ed.) (1986) *Competition in Global Industries* (Cambridge, Mass.: Harvard Business School Press).
PORTER, M. E. (1990) *The Competitive Advantage of Nations* (New York: The Free Press).
PRAHALAD, C. K. and Y. L. DOZ (1987) *The Multinational Mission: Balancing Local Demands and Global Vision* (New York: The Free Press).
PRAHALAD, C. K. and G. HAMEL (1990) 'The Core Competence of the Corporation', *Harvard Business Review*, vol. 68, no. 3, pp. 79–81.
PRAHALAD, C. K. and G. HAMEL (1994) 'Strategy as a Field of Study', *Strategic Management Journal*, Summer.
PRIME MINISTER'S OFFICE (1979) *Soshiki de Hataraku Seshonen no Ishiki* (The Attitude of Young Workers in the Organization) (Tokyo: Prime Minister's Office).
PUCIK, V., N. M. TICHY and C. K. BARLETT, (eds) (1993) *Globalizing Management* (New York: John Wiley).
QUELCH, J. A. and D. KENNY (1994) 'Extend Profits, Not Product Lines', *Harvard Business Review*, vol. 72, no. 5, pp. 153–60.
QUINN, J. B. (1978) 'Strategic Change? Logical Incrementalism', *Sloan Management Review*, Fall, pp. 7–21.

QUINN, J. B. (1980a) 'Managing Strategic Change', *Sloan Management Review*, no. 4, pp. 3–20.

QUINN, J. B. (1980b) *Strategies for Change: Logical Incrementalism* (Homewood, Ill.: Irwin).

QUINN, J. B., H. MINTZBERG and R. M. JAMES (1988) *The Strategy Process* (Englewood Cliffs, NJ: Prentice-Hall).

RAGER, R. K. and A. S. HUFF (1993) 'Strategic Groups, A Cognitive Perspective', *Strategic Management Journal*, vol. 14, no. 2, pp. 103–23.

RAO, A. R., M. E. BERGEN and S. DAVIS (2000) 'How to Fight a Price War', *Harvard Business Review*, vol. 78, no. 2, pp. 107–16.

RICHARD, M. D. (1978) *Organizational Goal Structures* (Minnesota: West Publishing).

RICHMAN, R. and M. COPEN (1972) *International Management and Economic Development* (New York: McGraw-Hill).

RICHTER, F. J. and K. VETTEL (1995) 'Successful Joint Ventures in Japan transferring knowledge through organizational learning', *Long Range Planning*, vol. 28, no. 3, pp. 37–45.

RICKS, D. A. (1983) *Big Business Blunders* (New York: Dow Jones Irwin).

RIESMAN, D. (1953) *The Lonely Crowd* (New Haven, CT: Yale University Press).

ROBIN, D., M. GIALLOURAKIS, F. David and T. MORITZ (1989) 'A Different Look at Codes of Ethics', *Business Horizons*, vol. 32, pp. 66–73.

ROBBINS, S. R. (1974) *Managing Organizational Conflict* (Englewood Cliffs, NJ: Prentice-Hall).

ROBINSON, E. A. G. (1958) *The Structure of Competitive Industry* (Chicago, Ill.: University of Chicago Press).

ROBOCK, S. K. and K. SIMMONDS (1989) *International Business and Multinational Enterprises*, 4th edn (Homewood, Ill.: Richard Irwin).

ROETHLISBERGER, F. J. and W. J. DICKSON (1939) *Management and the Worker* (Cambridge, Mass.: Harvard University Press).

ROGERS, E. M. (1932, 1962) *Diffusion of Innovations* (New York: The Free Press).

RONEN, S. (1986) *Comparative and Multinational Management* (New York: John Wiley).

ROTHBERG, R. R. (ed.) (1981) *Corporate Strategy and Product Innovation* (New York: The Free Press).

ROTHCHILD, W. E. (1976) *Putting It All Together: A Guide to Strategic Thinking* (New York: AMACOM).

ROTTER, J. B. (1995) 'Interpersonal Trust, Trustworthiness and Gullibility', *American Psychologist*, vol. 35, no. 1, pp. 1–12.

RUCE, R. and H. RAIFFA (1957) *Games and Decisions* (New York: John Wiley).

RUGMAN, A. M., D. J. LECREW and L. D. BOOTH (1985) *International Business* (New York: McGraw-Hill).

RUMELT, R. P. (1979) 'Evaluation of Strategy: Theory and Models', in D. E. SCHENDEL and C. HOFER (eds), Strategic Management (Boston, Mass.: Little, Brown), pp. 196–212.

RUMELT, R. (1984) 'The Evaluation of Business Strategy', in W. F. GLUECK (ed.), *Business Policy and Strategic Management* (New York: McGraw-Hill).

SALANCIK, G. R. and J. PFEFFER (1977) 'Who Gets Power and How They Hold on to It, A Strategic Contingency Model of Power, in *Organizational Dynamics* (New York: ANACOM).

SASAKI, T. (1993) 'Strategic Alliances in the Car Industry', *Long Range Planning*, October.

SENNETT, R. (1999) *The Corrosion of Character* (Oxford: Polity Press).

SCHEIN, E. H. (1985) *Organizational Culture and Leadership* (San Francisco, CA: Jossey-Bass).

SCHMIEGELOW, M. (1989) *Strategic Pragmatism – Japanese Lessons in the Use of Economic Theory* (Bonn: Praeger).

SCHOFIELD, M. and D. ARNOLD (1988) 'Strategies for Mature Business', *Long Range Planning*, October, pp. 69–76.

SCHOLLHAMMER, H. (1969) 'The Comparative Management Theory Jungle', *Academy of Management Journal*, vol.12 (March), pp. 81–97.

SCHONBERGER, R. J. (1982) 'The Transfer of Japanese Manufacturing Management Approaches to U.S. Industry', *Academy of Management Review*, vol. 7, pp. 479–87.

SCHULER, R. S. and S. E. JACKSON (1987) 'Linking Competitive Strategies with Human Resource Management Practices', *Academy of Management Executive*, vol. 1, no.3, pp. 207–19.

SCOTT, B. R. (1971) *Stages of Corporate Development* (Boston, Mass.: Harvard Business School).

SHARPE, W. F. (1970) *Portfolio Theory and Capital Markets* (New York: McGraw-Hill).

SHAUGHNESSY, H. (1995) 'International Joint Ventures: Managing Successful Collaborations', *Long Range Planning*, vol. 28, no.3, pp. 10–17.

SHIBAKAWA, R. (1987) *Nihon Kigyo no Keiei Kokusai-ka* (Globalization of Japanese Enterprises) (Tokyo: Chuokeizai).

SHIBATA, G. and K. NAKAHASHI (1997) *Keiei Kanri no Riron to Jissen* (Theory and Practice of Management) (Tokyo: Joho Shuppan).

SHIMOTANI, M. (1993) *Nihon no Keiretsu to Kigyo Group* (Japanese Keiretsu and Business Groups) (Tokyo: Yuhikaku).

SIMON, H. A. (1957) *Administrative Behavior* (New York: Macmillan).

SMITH, P. (1997) *Japan: A Reinterpretation* (New York: Pantheon).

SOMUCHO (General Affairs Bureau) (1989) *Sekai Seinen Ishiki* (Survey on Attitude of World Youth) (Tokyo: Somucho).

STABELL, C. B. and O. D. FILEDSTAD (1998) 'Configuring Value for Competitive Advantage: On Chains, Shops and Networks', *Strategic Management Journal*, vol. 19, no.5, pp. 413–37.

STACEY, R. D. (1995) 'The Science of Complexity: An Alternative Perspective for Strategic Change Processes', *Strategic Management Journal*, vol. 16, no.6, pp. 477–95.

STALK, G., P. EVANS and L. E. SHULMAN (1992) 'Competing on Capabilities, the new rules of corporate strategy', *Harvard Business Review*, vol. 70, no.2, pp. 57–69.

STAW, B. M. (1976) *Intrinsic and Extrinsic Motivation* (New Jersey: General Learning Press).
STEINER, C. (ed.) (1965) *The Creative Organization* (Chicago, Ill.: University of Chicago Press).
STEINER, G. A. (1969) *Top Management Planning* (New York: Macmillan).
STEINER, G. A. and J. B. MINER (1977) *Management Policy and Strategy* (New York: Macmillan).
STEWART, G. B. (1991) *The Quest For Value – The EVA Management Guide* (New York: HarperCollins).
STOBAUGH, R. (1972) 'How Investment Abroad Create Jobs at Home', *Harvard Business Review*, vol. 50, pp. 118–260.
STOPFORD, J. M. and L. T. WELLS (1972) *Managing the Multinational Enterprise* (New York: Basic Books).
SUGIURA, H. (1990) 'How Honda Localized Its Global Strategy', *Sloan Management Review*, Fall, pp. 77–82.
SUZUKI, M. (1959) *Game no Ribon* (Theory of Games) (Tokyo: Keiso Shobo).
SUZUKI, N. (1988) *Takokuseki Kigyo Keieiron* (Study of Multinational Management) (Tokyo: Kokusai Shobo).
TAKAHASHI, T. (ed.) (1995) *Corporate Governance* (Tokyo: Chuokeizai).
TAKAHASHI, T. and T. ONISHI (eds) (1997) *Doitsu no Kigyo* (The German Corporation) (Waseda: Daigaku Shuppan).
TANNO, I. (1993) *Kokusai Hikaku Keieiron* (Comparative Management) (Tokyo: Dobunkan).
TERRY, R. J. (1982) 'Ten Suggestion for Acquisition Success', *Managerial Planning*, September–October.
THOMAS, R. J. (1995) *New Product Success Stories* (New York: John Wiley).
THOMPSON JN, A. A. STRICKLAND III, A. J. and KRAMER, T. R. (1993) *Readings in Strategic Management* (Homewood, Ill.: Irwin).
THOMPSON, J. (1967) *Organizations in Action* (New York: McGraw-Hill).
TILLES, S. (1963) 'How to Evaluate Strategy', *Harvard Business Review*, July–August, pp. 111–21.
TONNIES, F. (1987/1963) *Gemeinschaft and Gesselschaft* (Community and Society) (New York: Harper & Row).
TSUCHIYA, M. (1984) *Kigyo to Senryaku* (Corporation and Strategy) (Tokyo: Recruit).
TSUCHIYA, M. and Y. KONOMI (1995) *Korekara no Nihon-teki Keiei* (New Japanese Management) (Tokyo: Nihonhoso Suppan).
TSUCHIYA, T. (1964) *Nihon Keiei Rinenshi* (History of Japanese Corporate Philosophy) (Tokyo: Nihonkeizai).
TSUCHIYA, T. (1967) *Zoku Nihon Keiei Rinenshi* (History of Japanese Corporate Philosophy–Continued) (Tokyo: Nihonkeizai).
TOYO KEIZAI (1993)
TSUCHIYA, T. (1984) *TQC* (Tokyo: Nikkeiren).
TSUTSUI, W. M. (1998) *Manufacturing Ideology: Scientific Management in Twentieth-Century Japan* (Princeton, NJ: Princeton University Press).
TYLER, B. B. and H. K. STEENSMA (1995) 'Evaluating Technological Collaborative Opportunities', *Strategic Management Journal*, Summer, pp. 43–70.

UCHIHASHI, K. (1978) *Takumi no Jidai* (Inside Story of New Products Development) (Tokyo: Sankei Shuppan).

UEKI, H. (1982) *Kokusai Keiei Shigen Henron* (International Transfer of Business Resources) (Tokyo: Bunshodo).

UETANI, A. (1997) *ISO14001, System Ko-chiku Ninsho Guide* (ISO14001, System Construction and Lincense Guide) (Tokyo: Nikkei BP).

URABE, K. (1984) *Nihon teki Keiei wa Shinka Suru* (Progress of the Japanese Management System) (Tokyo: Chuo Keizai).

URBAN, G. L. and HAUSER, J. R. (1987) *Essentials of New Product Management*, (Englewood Cliffs, NJ: Prentice-Hall).

URBAN, G. L. and J. R. HAUSER (1993) *Design and Marketing of New Products* (Englewood Cliffs, NJ: Prentice-Hall).

VERNON, R. (1966) 'International Investment and International Trade in the Product Cycle', *The Quarterly Journal of Economics*, May.

VERNON, R. (1971) *Sovereignty at Bay, The Multinational Spread of US Enterprises* (New York: Basic Books).

VOGEL, E. F. (1979) *Japan as Number One: Lessons for America* (Cambridge, Mass.: Harvard University Press).

VON, H. E. (1988) *The Sources of Innovation* (New York: Oxford University Press).

WALTON, R. F. (1985) 'From Control to Commitment in the Workplace', *Harvard Business Review*, March–April, pp. 81–3.

WATHNE, K., J. ROSS and G. KROGH (1995) 'The Impact of Trust and Experience on Knowledge Transfer in a Corporative Context', paper presented at the Strategic Management Society Conference, Mexico City.

WATSON, G. H. (1992) *The Benchmarking Workbook* (Portland, Oregon: Productivity Press).

WEBER, M. (1904) *The Protestant Ethic and the Spirit of Capitalism* (New York: Charles Scribner's sons).

WEICK, K. E. (1979) *The Social Psychology of Organizing*, 2nd edn (Reading, Mass.: Addison-Wesley).

WEST, M. A. and FARR, J. L. (1990) *Innovation and Creativity at Work* (London: John Wiley).

WESTNEY, D. E. (1987) *Initiation and Innovation: The Transfer of Western Organizational Patterns to Meiji Japan* (Cambridge, Mass.: Harvard University Press).

WHEELRIGHT, S. C. (1981) 'Japan: Where Operations Really are Strategic', *Harvard Business Review*, vol. 59, no.4, pp. 67–74.

WHEELWRIGHT, S. C. and K. B. CLARK (1992) *Revolutionizing Product Development* (New York: Free Press).

WHITEHILL, L. A. (1991) *Japanese Management* (London: Routledge).

WHYTE, W. H. (1956) *The Organization Man* (New York: Simon and Schuster).

WIERSEMA, M. F. and K. A. BANTEL (1993) 'Top Management Team Turnover as an Adaptation Mechanism, The Role of the Environment', *Strategic Management Journal*, vol. 14, no.7, pp. 485–504.

WILLIAMSON, O. (1975) *Markets and Hierarchies* (New York: Free Press).

WILSON, I. (1992) 'Realizing the Power of Strategic Vision', *Long Range Planning*, vol. 25, no.5, pp. 18–27.

WOOD, C. (1994) *The End of Japan Inc.* (New York: Simon & Schuster).
WOODWARD, J. (1965) *Industrial Organization, Theory and Practice* (Oxford: Oxford University Press).
WORTZEL, H. V. and L. H. WORTZEL (eds) (1991) *Global Strategic Management* (New York: John Wiley).
YAMANOUCHI, A. (1986) *Kigyou Henkaku no Gijutsu Management* (Management of Technology for Innovation) (Tokyo: Nihonkeizai).
YAMANOUCHI, A. (1989) 'Breakthrough: The Development of the Canon Personal Copier', *Long Range Planning*, vol. 22, no.5, pp. 11–21.
YAMANOUCHI, A. (1992) *Shin Gijitsu Keiei Ron* (New Research Management) (Tokyo: Nihonkeizai).
YAMASHIRO, A. (1967) *Gendai no Keiei Rinen* (Modern Business Creed) (Tokyo: Hakuto Shobo).
YAMASHIRO, A. (ed.) (1969) *Gendai no Keiei Rinen [Riron Hen]* (Business Philosophy [Theories]) (Tokyo: Hakuto Shobo).
YASUMURO, K. (1992) *Global Keieiron* (Global Management) (Tokyo: Chikura Shobo).
YIP, G. S. (1989) *Total Global Strategy* (Englewood Cliffs, NJ: Prentice-Hall).
YOSHIDA, K. (1993) *Nihongata Keiei System no Kozai* (Merits and Demerits of the Japanese Management System) (Tokyo: Toyokeizai).
YOSHIDA, K. (1996) *Nihongata Keiei System* (Japanese Management System) (Tokyo: Toyo Keizai Shinpo-sha).
YOSHIHARA, H. (1977) 'The Japanese Multinational', *Long Range Planning*, vol. 10, no.2, pp. 41–5.
YOSHIHARA, H. (1986) *Senryakuteki Kigyo Kakushin* (Strategic Innovation), (Tokyo: Toyokeizai).
YOSHIHARA, E., Y. HAYASHI and K. YASUMURO (1988) *Nuhon Kigyo no Global Keiei* (Global Management of Japanese Firms) (Tokyo: Toyokeizai).
YOSHIMORI, K. (1995a) 'Doitsu no Keieisha' (German Top Management), in T. Takahashi (ed.), *Corporate Governance* (Tokyo: Chuokeizai).
YOSHIMORI, K. (1995b) 'The Concept of the Corporation in Japan and the West', *Long Range Planning*, vol. 28, no. 4, pp. 33–44.
YOSHIMORI, K. (1997) 'Kigyo Tochi' (Corporate Governance), Report at a study meeting of the Japanese Academy of Management.
YOSHIMURA, N. and P. ANDERSON (1997) *Inside Kaisha: Demystifying Japanese Business Behavior* (Boston, Mass.: Harvard Business School Press).
YOSHINO, M. Y. (1968) *Japan's Managerial System: Tradition and Innovation* (Cambridge, Mass.: MIT Press).
YOUNDT, M. A., S. A. SNELL, J. W. DEAN and D. P. LEPAK (1996) 'Human Resource Management, Manufacturing Strategy and Firm Performance', *Academy of Management Journal*, vol. 39, no.4, pp. 836–66.

Index